George Frederick Childe

Singular properties of the Ellipsoid

and associated surfaces of the Nth degree

George Frederick Childe

Singular properties of the Ellipsoid
and associated surfaces of the Nth degree

ISBN/EAN: 9783744650625

Printed in Europe, USA, Canada, Australia, Japan

Cover: Foto ©berggeist007 / pixelio.de

More available books at **www.hansebooks.com**

SINGULAR PROPERTIES

OF THE

ELLIPSOID,

AND

ASSOCIATED SURFACES OF THE Nᵀᴴ DEGREE.

DEDICATED, BY PERMISSION,

TO

HIS ROYAL HIGHNESS PRINCE ALFRED.

BY THE REV. G. F. CHILDE, M.A.,

AUTHOR OF RAY SURFACES, RELATED CAUSTICS, &c.
MATHEMATICAL PROFESSOR IN THE SOUTH AFRICAN COLLEGE.
MEMBER OF THE BOARD OF PUBLIC EXAMINERS AT THE CAPE OF GOOD HOPE.

MACMILLAN AND Co., CAMBRIDGE; J. C. JUTA, CAPE TOWN.

MDCCCLXI.

CAPE TOWN:

W. F. MATHEW, STEAM PRINTING OFFICE,

ST. GEORGE STREET.

TO HIS ROYAL HIGHNESS

PRINCE ALFRED,

IN REMEMBRANCE OF HIS VISIT TO THE COLONY OF THE

CAPE OF GOOD HOPE, IN THE YEAR 1860,

AND OF HIS PERMISSION THEN GRACIOUSLY GIVEN,

THIS VOLUME IS DEDICATED,

BY HIS FAITHFUL, HUMBLE SERVANT,

THE AUTHOR.

INTRODUCTORY REMARKS.

As the title of this volume indicates, its object is to develope peculiarities in the Ellipsoid; and, further, to establish analogous properties in the unlimited congeneric series of which this remarkable surface is a constituent.

The more conspicuous Singularities which have been evolved are enumerated in the Table of Contents; among them it seems desirable to specify more pointedly, and as briefly as possible, those which follow.

(1.)—The relation of the Circle-ordinate u to the co-ordinates. This relation is first noticed in (6), p. 2; and will be found to pervade the whole subsequent investigation. (2.)—The equation (13), p. 4; and the cognate relations developed throughout Chapter IV. (3.)—The series of equations exhibited in page 8. (4.)—The very peculiar relation of the intercept ρ_m to the co-ordinates, investigated for the Ellipsoid in (4), p. 10; and, for the General Surface, in (4), p. 22. 5.—The summary of General Formulæ commencing at p. 23. In the Table of Errata it is pointed out that the forms (5) (8) (9) are erroneous, and the requisite corrections have been there introduced. 6.—The remarkable system of equations representing successive planes in the ellipsoid, p. 30; and tangent-planes, p. 33. 7.—The developments of Chapter III. The geometrical peculiarities detected by means of these developments, pp. 51 and 56, deserve to be particularised. (8)—The singular series of expressions for the tangent perpendicular in consecutive surfaces. (9.)—Properties of the Circle-ordinates in pp. 97 and 102. (10.)—The axes of the mth derived ellipsoid, p. 109. (11.)—Singular relations involving the Circle-ordinates in consecutive surfaces, p. 113. (12.)—The singular relations combining consecutive co-ordinates and constants, in different surfaces, p. 125. (13.)—The relation uniting the Circle-ordinates in any three consecutive surfaces (XXXII), p. 130. (18.)—The singular equation connecting successive Circle-ordinates in the general surface, p. 132.

In Chapter VI, relating to the Ellipsoid, the area of a plane section, which is determined under the most general circumstances, has a singular symmetry of expression. It is possible that this result may be otherwise obtained in some more compendious way, although no method more convenient has occurred to the author. The identification of the area with a second orthographical projection of the parallel central section, given at page 140, illustrates the clearness of interpretation which may attach to results of a purely analytical character, by reference to their geometrical equivalence. In applying the expression for a sectional area to the Volume of any portion of the solid limited by parallel planes, we are led to results of no ordinary simplicity regarding the volume or mass; while the extension of this investigation in determining the attractive force, under the hypothesis that the attraction varies directly with the distance, may not be without its value, in the consideration of problems relating to this subject.

With reference to this point, it should be remarked that, in the formula (2), page 148, the Density may be assumed to be any function of the distance, although, in the subsequent propositions, it has been taken as constant. That illustrations in greater detail, as well on this, as on some other topics, have not been given, must be attributed to the narrow limits within which the writer has been unavoidably restricted.

If it is of acknowledged difficulty in the present day, to bring forward anything of novelty in mathematical researches, yet it may be hoped that such attempts are not often altogether futile, when entered upon in the spirit which is anxious for the investigation of Truth, and desires to add its contribution to the treasury of Science.

To do this, in some degree, is the design of the investigations exhibited in the following pages. For whatever defect in its execution may be apparent, the reader's indulgence is requested; while, for the recognition of whatever truth has been elicited, the author is content to rest upon the candour of an impartial judgment.

TABLE OF CONTENTS.

CHAPTER I.

CHAPTER II.

* NOTE.—The formulæ (5) (8) (9), at pages 24 and 25, which are inaccurate in the text, have been corrected in the table of errata.

viii CONTENTS.

CHAPTER III.

CHAPTER VI.

TABLE OF ERRATA.

PAGE.

8, (VII); 19, (IV); 21, (V). For v *read* v_m.

10, line 5. For (VI) *read* (VII).

11, (IX). For $v_{/}$ *read* $v_{,m}$.

24, (VII). In form (5), for Σ^2, in the *numerator*, *read* $\Sigma^{2(n-1)}$.

25, (VII). Instead of form (8), *substitute*,

$$\frac{p^2_m}{r^2_m} = \frac{\Sigma^3 \left\{ x^{\frac{(n-1)}{\frac{m}{n-2}}} \cdot \frac{1}{a^{\frac{n(n-1)-2}{n-2}}} \right\}^n}{\Sigma \left\{ \frac{x^{\frac{(n-1)}{\frac{m+1}{n-2}}}}{a^{n\frac{(n-1)-1}{n-2}}} \right\}^2 \Sigma \left\{ \frac{x^{\frac{(n-1)}{\frac{m}{n-2}}}}{a^{n\frac{(n-1)-1}{n-2}}} \right\}^2}.$$

25, (VII), In form (9), for Σ in the *numerator*, *read* Σ^{n-1}.

33, (XIII). For v *read* v_m.

34, line 1. After page 25 *read* (corrected).

42, line 4. For a^9 *read* x^9.

43, line 1. For a^{12} *read* x^{12}.

65. In (8) (9) (10), *invert* $\frac{x}{a}$ under Σ.

67. For the NOTE, *read*,
    "See Gregory's General Theorems in the Calculus. Professor Kelland
    on General Differentiation (Edinburgh Phil. Transactions), &c."

73. After (3) *insert* $\}$

73, line 9. For p *read* r.

103, line 19. In c_2 for b *read* v.

135, line 11. For $x^2 y^3$ *read* $x_2 \, y_2$.

147, line 7. For *fraction* read *function*.

SINGULAR PROPERTIES

OF THE

ELLIPSOID.

CHAPTER I.

(I.) In the following pages it is proposed to investigate some singular relations which exist between lines connected with the Ellipsoid, and with the surface represented by the general equation,

$$\frac{x^n}{a^n} + \frac{y^n}{b^n} + \frac{z^n}{c^n} = 1.$$

The properties of the ellipsoid will be considered in the first instance, the examination being afterwards extended to the general surface.

Let us suppose a tangent plane to be drawn at any point P on the ellipsoid (fig. 1), O being the centre; let $OP = r$ be the radius vector drawn to the point of contact from the centre; $Op = p$ the perpendicular from the centre upon the tangent plane; $OP_{,} = r_{,}$ the radius vector intercepted upon this perpendicular between the surface and its centre; we may then shew that these three lines are connected by the equation,

$$r^2 p^2 r_{,}^2 - (a^2 + b^2 + c^2) p^2 r_{,}^2 + (a^2 b^2 + a^2 c^2 + b^2 c^2) r_{,}^2 - a^2 b^2 c^2 = 0:$$

in which $a\,b\,c$ are the semiaxes of the surface.

(II.) Let $\xi\,\eta\,\zeta$ be the general co-ordinates of the tangent-plane, $x\,y\,z$ any point on the ellipsoid; these two surfaces are defined by the equations,

$$\frac{x^2}{a^2} + \frac{y^2}{b^2} + \frac{z^2}{c^2} = 1. \tag{1}$$

$$\frac{x\xi}{a^2} + \frac{y\eta}{b^2} + \frac{z\zeta}{c^2} = 1. \tag{2}$$

B

The equations of the perpendicular from the centre upon the tangent plane are,

$$a^2 z\xi - c^2 x\zeta = 0 \atop b^2 z\eta - c^2 y\zeta = 0 \Big\}. \tag{3}$$

By combining (2) and (3) we obtain the usual expression for p,

$$\frac{1}{p^2} = \frac{x^2}{a^4} + \frac{y^2}{b^4} + \frac{z^2}{c^4}. \tag{4}$$

Let x_i, y_i, z_i be co-ordinates of the point in which p intersects the surface, then for the determination of these quantities there are the equations,

$$\frac{x_i^2}{a^2} + \frac{y_i^2}{b^2} + \frac{z_i^2}{c^2} = 1$$

$$a^2 z x_i - c^2 x z_i = 0; \quad b^2 z y_i - c^2 y z_i = 0;$$

from which, after assuming

$$\frac{x^2}{a^6} + \frac{y^2}{b^6} + \frac{z^2}{c^6} = \frac{1}{u^4}, \tag{5}$$

we find as the values of the co-ordinates of intersection,

$$x_i = \pm \left(\frac{u}{a}\right)^2 x. \quad y_i = \pm \left(\frac{u}{b}\right)^2 y. \quad z_i = \pm \left(\frac{u}{c}\right)^2 z.$$

Now,
$$r_i^2 = x_i^2 + y_i^2 + z_i^2$$
$$= u^4 \left(\frac{x^2}{a^4} + \frac{y^2}{b^4} + \frac{z^2}{c^4}\right)$$
$$= \frac{u^4}{p^2}$$
$$\therefore \quad u^2 = p r_i; \tag{6}$$

from which it appears that the line represented by u is a mean proportional to p and r_i.

Hence we have, for the elimination of $x\, y\, z$, the four equations,

$$\frac{x^2}{a^2} + \frac{y^2}{b^2} + \frac{z^2}{c^2} = 1 \tag{7}$$

$$\frac{x^2}{a^4} + \frac{y^2}{b^4} + \frac{z^2}{c^4} = \frac{1}{p^2} \tag{8}$$

$$\frac{x^3}{a^6} + \frac{y^3}{b^6} + \frac{z^2}{c^6} = \frac{1}{p^2 r_{\prime}^{~2}} \qquad (9)$$

$$x^2 + y^2 + z^2 = r^3 \qquad (10)$$

It should be here remarked that the auxiliary symbol u, employed in the reduction, has the following geometrical signification.

Let the radius $OP_{\prime}$ be produced to meet the surface on the opposite side in the point Q; $P_{\prime}OQ$ is then a diameter, and $OQ = r_{\prime}$. On Qp describe a semicircle in the plane $OPP_{\prime}$, and draw the ordinate OL: then,

$$OL^2 = OQ \times Op = pr_{\prime}.$$

Consequently, u is equal to the line OL; which it is obvious might equally well be taken as the ordinate of any other circle, or a semichord of the sphere, described on Qp as diameter.

(III.) In eliminating $x\, y\, z$ from the four equations last written, it is convenient to adopt the notation,

$$a_m = \frac{1}{a^{2m}}; \qquad \beta_m = \frac{1}{b^{2m}}; \qquad \gamma_m = \frac{1}{c^{2m}};$$

so that the suffixed symbols shall be subject to the law of indices; e.g. $a_m\, a_n = a_{m+n}$: then,

$$a x^2 + \beta y^2 + \gamma z^2 = 1 \qquad (7)$$

$$a_2 x^2 + \beta_2 y^2 + \gamma_2 z^2 = P \qquad (8)$$

$$a_3 x^2 + \beta_3 y^2 + \gamma_3 z^2 = Q \qquad (9)$$

writing, $\qquad \dfrac{1}{p^2} = P; \qquad \dfrac{1}{p^2 r_{\prime}^{~2}} = Q.$

By employing indeterminate multipliers, we shall find,

$$\left. \begin{aligned} x^2 &= \frac{\beta\gamma - (\beta + \gamma)\, P + Q}{a\, (a - \beta)\, (a - \gamma)} \\[1em] y^3 &= \frac{-a\gamma + (a + \gamma)\, P - Q}{\beta\, (a - \beta)\, (\beta - \gamma)} \\[1em] z^3 &= \frac{a\beta - (a + \beta)\, P + Q}{\gamma\, (a - \gamma)\, (\beta - \gamma)} \end{aligned} \right\} \qquad (11)$$

Now, when $x\ y\ z$ are eliminated from the equation (10), there will be found,

$$r^2 \quad = \quad A \quad - \quad BP \quad + \quad CQ; \tag{12}$$

if we make,

$$A = \frac{a\beta}{\gamma\,(a-\gamma)\,(\beta-\gamma)} - \frac{a\gamma}{\beta(a-\beta)(\beta-\gamma)} + \frac{\beta\gamma}{a(a-\beta)\,(a-\gamma)} = a^2 + b^3 + c^2.$$

$$B = \frac{a+\beta}{\gamma\,(a-\gamma)\,(\beta-\gamma)} - \frac{a+\gamma}{\beta(a-\beta)(\beta-\gamma)} + \frac{\beta+\gamma}{a\,(a-\beta)(a-\gamma)} = a^2b^2 + a^2c^2 + b^2c^2.$$

$$C = \frac{1}{\gamma(a-\gamma)\,(\beta-\gamma)} - \frac{1}{\beta(a-\beta)(\beta-\gamma)} + \frac{1}{a(a-\beta)\,(a-\gamma)} = a^2b^2c^2.$$

Hence the relation between rpr, becomes, finally,

$$r^2p^2r_{\prime}^2 - (a^2+b^2+c^2)\,p^2r_{\prime}^2 + (a^2b^2+a^2c^2+b^2c^2)\,r_{\prime}^2 - a^2b^2c^2 = 0 \tag{13}$$

Cor. 1. If we suppose $r = p = r$, this equation resolves itself into,

$$(r^2 - a^2)\,(r^2 - b^2)\,(r^2 - c^2) \quad = \quad 0\,;$$

shewing that the assumed equality can exist only on the principal axes of the surface, as we know to be the case à priori: and if $a = b = c$, the surface being spherical, this equation is reduced to

$$(r^2 - a^2)^3 \quad = \quad 0.$$

If any one of the axes is supposed to be evanescent, r, disappears from the equation, which becomes that of the corresponding plane elliptic section. For example, let $c = o$, then,

$$(a^2 + b^2 - r^2)\,p^2 = a^2b^2.$$

Cor. 2. It is to be observed that the co-efficients and indices in (13) follow the law of combination which determines those of a complete cubic equation.

Cor. 3. When any axis of the ellipsoid is supposed to become infinite, that surface passes into an elliptic cylinder, and the equation (13) ought to exhibit the relation between p and r, in the plane ellipse. Writing $c = \infty$, we find,

$$(a^2 + b^2 - p^2)\, r_i^{\,2} = a^2 b^2\,;$$

which will appear, by an independent elimination, to be the proper relation in the ellipse defined by,

$$\frac{x^2}{a^2} + \frac{y^2}{b^2} = 1.$$

Cor. 4. A curious analogy exists between the foregoing form and the relation between p and r in the same ellipse, which is,

$$(a^2 + b^2 - r^2)\, p^2 = a^2 b^2.$$

By a comparison of the two equations it will be perceived that, if a point is taken on the ellipse at which the radius vector is equal to any tangent-perpendicular, the intercept on the first perpendicular will be always equal to the tangent-perpendicular corresponding to that point.

(IV.) Let us now suppose a series of points on the surface, whose positions derive from one another consecutively, in the same manner as P_i is determined by P ; that is to say, P_m is a point in which the surface is cut by p_{m-1} the perpendicular upon the tangent plane of P_{m-1} : r_m is the intercept upon p_{m-1}, or the radius vector of P_m : $x_m\, y_m\, z_m$ the rectangular co-ordinates of P_m : we have then the following series of equalities,

$$\frac{1}{p^2} = \frac{x^2}{a^4} + \frac{y^2}{b^4} + \frac{z^2}{c^4}.$$

$$\frac{1}{p_i^{\,2}} = \frac{x_i^{\,2}}{a^4} + \frac{y_i^{\,2}}{b^4} + \frac{z_i^{\,2}}{c^4}.$$

$$\cdots \qquad \cdots \qquad \cdots \qquad \cdots$$

$$\frac{1}{p_m^{\,2}} = \frac{x_m^{\,2}}{a^4} + \frac{y_m^{\,2}}{b^4} + \frac{z_m^{\,2}}{c^4}.$$

Again, considering the lines analogous to u which correspond to the consecutive points, we have the series,

$$\frac{1}{u^4} = \frac{x^2}{a^6} + \frac{y^2}{b^6} + \frac{z^2}{c^6}.$$

$$\frac{1}{u_i^{\,4}} = \frac{x_i^{\,2}}{a^6} + \frac{y_i^{\,2}}{b^6} + \frac{z_i^{\,2}}{c^6}.$$

$$\cdots \qquad \cdots \qquad \cdots \qquad \cdots$$

$$\frac{1}{u_m^{\,4}} = \frac{x_m^{\,2}}{a^6} + \frac{y_m^{\,2}}{b^6} + \frac{z_m^{\,2}}{c^6}.$$

But, $x_{\prime} = \left(\dfrac{u}{a}\right)^2 x$; $y_{\prime} = \left(\dfrac{u}{b}\right)^2 y$; $z_{\prime} = \left(\dfrac{u}{c}\right)^2 z$;

and, as the law of derivation is uniform throughout the series,

$$x_2 = \left(\dfrac{u_{\prime}}{a}\right)^2 x_{\prime}; \quad y_2 = \left(\dfrac{u_{\prime}}{b}\right)^2 y_{\prime}; \quad z_2 = \left(\dfrac{u_{\prime}}{c}\right)^2 z_{\prime};$$

$$\dots \qquad \dots \qquad \dots \qquad \dots \qquad \dots \qquad \dots$$

$$x_m = \left(\dfrac{u_{m-1}}{a}\right)^2 x_{m-1}; \quad y_m = \left(\dfrac{u_{m-1}}{b}\right)^2 y_{m-1}; \quad z_m = \left(\dfrac{u_{m-1}}{c}\right)^2 z_{m-1};$$

hence will be obtained the general expressions,

$$x_m = \left(\dfrac{u_{m-1}}{a}\right)^2 \left(\dfrac{u_{m-2}}{a}\right)^2 \dots \left(\dfrac{u}{a}\right)^2 x.$$

$$y_m = \left(\dfrac{u_{m-1}}{b}\right)^2 \left(\dfrac{u_{m-2}}{b}\right)^2 \dots \left(\dfrac{u}{b}\right)^2 y.$$

$$z_m = \left(\dfrac{u_{m-1}}{c}\right)^2 \left(\dfrac{u_{m-2}}{c}\right)^2 \dots \left(\dfrac{u}{c}\right)^2 z.$$

Cor. The ratios of the mth co-ordinates vary as the ratios of the first co-ordinates; for,

$$\dfrac{x_m}{y_m} = \left(\dfrac{b}{a}\right)^{2m} \dfrac{x}{y}; \quad \dfrac{x_m}{z_m} = \left(\dfrac{c}{a}\right)^{2m} \dfrac{x}{z}; \quad \dfrac{y_m}{z_m} = \left(\dfrac{c}{b}\right)^{2m} \dfrac{y}{z};$$

or, $x_m \ : \ y_m \ : \ z_m \ :: \ (bc)^{2m} x \ : \ (ac)^{2m} y \ : \ (ab)^{2m} z.$

(V.) We have next to calculate $p_{\prime}^2 \, p_2^2 \dots u_{\prime}^2 \, u_2^2 \dots$, employing the formulæ of the last article;

$$\dfrac{1}{p_{\prime}^2} = \dfrac{x_{\prime}^2}{a^4} + \dfrac{y_{\prime}^2}{b^4} + \dfrac{z_{\prime}^2}{c^4}$$

$$\therefore \ \dfrac{1}{p_{\prime}^2} = u^4 \left\{ \dfrac{x^2}{a^8} + \dfrac{y^2}{b^8} + \dfrac{z^2}{c^8} \right\}$$

similarly, $\dfrac{1}{p_2^2} = u_{\prime}^4 \, u^4 \left\{ \dfrac{x^2}{a^{12}} + \dfrac{y^2}{b^{12}} + \dfrac{z^2}{c^{12}} \right\}$

$$\dots \qquad\qquad \dots$$

and, generally,

$$\dfrac{1}{p_m^2} = u_{m-1}^4 \, u_{m-2}^4 \dots u^4 \left\{ \dfrac{x^2}{a^{4m+4}} + \dfrac{y^2}{b^{4m+4}} + \dfrac{z^2}{c^{4m+4}} \right\}.$$

Again,
$$\frac{1}{u_{\prime}^{4}} = \frac{x_{\prime}^{2}}{a^{6}} + \frac{y_{\prime}^{2}}{b^{6}} + \frac{z_{\prime}^{2}}{c^{6}},$$

$$\therefore \quad \frac{1}{u_{\prime}^{4}} = u^{4}\left\{\frac{x^{2}}{a^{10}} + \frac{y^{2}}{b^{10}} + \frac{z^{2}}{c^{10}}\right\};$$

and, generally,

$$\frac{1}{u_{m}^{4}} = u_{m-1}^{4}\, u_{m-2}^{4} \cdots u^{4}\left\{\frac{x^{2}}{a^{4m+6}} + \frac{y^{2}}{b^{4m+6}} + \frac{z^{2}}{c^{4m+6}}\right\}.$$

Now it has been shewn that $u^{2} = pr_{\prime}$; and, by the uniformity of the law of derivation we shall have, consequently,

$$u_{\prime}^{2} = p_{\prime}\, r_{2}\; ; \quad u_{2}^{2} = p_{2}\, r_{3}\; ; \quad u_{3}^{2} = p_{3}\, r_{4}\; ; \quad \ldots$$

$$u_{m}^{2} = p_{m}\, r_{m+1}.$$

Employing these equivalents in the reduction of the preceding expressions, and writing Σ to indicate the sum of three analogous quantities, we obtain,

$$\frac{1}{p_{\prime}^{2}} = p^{2}r_{\prime}^{2}\, \Sigma\left(\frac{x^{2}}{a^{6}}\right)$$

$$\frac{1}{p_{2}^{2}} = p_{\prime}^{2}\, r_{2}^{2}\, p^{2}\, r_{\prime}^{2}\; \Sigma\left(\frac{x^{2}}{a^{12}}\right)$$

$$\frac{1}{p_{3}^{2}} = p_{2}^{2}\, r_{3}^{2}\, p_{\prime}^{2}\, r_{2}^{2}\, p^{2}\, r_{\prime}^{2}\; \Sigma\left(\frac{x^{2}}{a^{16}}\right)$$

$$\cdots$$

$$\frac{1}{p_{m}^{2}} = p_{m-1}^{2}\, r_{m}^{2} \ldots p^{2}\, r_{\prime}^{2}\; \Sigma\left(\frac{x^{2}}{a^{4m+4}}\right).$$

And,
$$\frac{1}{p^{2}r_{\prime}^{2}} = \Sigma\left(\frac{x^{2}}{a^{6}}\right)$$

$$\frac{1}{p_{\prime}^{2}r_{2}^{2}} = p^{2}\, r_{\prime}^{2}\, \Sigma\left(\frac{x^{2}}{a^{10}}\right)$$

$$\frac{1}{p_{2}^{2}r_{3}^{2}} = p_{\prime}^{2}\, r_{2}^{2}\, p^{2}\, r_{\prime}^{2}\, \Sigma\left(\frac{x^{2}}{a^{14}}\right)$$

$$\cdots \qquad \cdots$$

$$\frac{1}{p_{m}^{2}r_{m+1}^{2}} = p_{m-1}^{2}\, r_{m}^{2}\, p_{m-2}^{2}\, r_{m-1}^{2} \ldots p^{2}\, r_{\prime}^{2}\, \Sigma\left(\frac{x^{2}}{a^{4m+6}}\right).$$

(VI.) After the preceding results have been arranged in order, we find that the following remarkable series of equations is derived from the co-ordinates of any point $x\,y\,z$ in the ellipsoid; according to the law previously stated :

$$\Sigma \left(\frac{x^2}{a^2}\right) = 1$$

$$\Sigma \left(\frac{x^2}{a^4}\right) = \frac{1}{p^2}$$

$$\Sigma \left(\frac{x^2}{a^6}\right) = \frac{1}{p^2 r_{\prime}^2}$$

$$\Sigma \left(\frac{x^2}{a^8}\right) = \frac{1}{p^2 p_{\prime}^2 r_{\prime}^2}$$

$$\cdots \qquad \cdots$$

$$\Sigma \left(\frac{r^2}{a^{4m+4}}\right) = \frac{1}{p_m^2\, p_{m-1}^2 \ldots p^2\, r_m^2\, r_{m-1}^2 \ldots r_{\prime}^2}.$$

$$\Sigma \left(\frac{x^2}{a^{4m+6}}\right) = \frac{1}{p_m^2\, p_{m-1}^2 \ldots p^2\, r_{m+1}^2\, r_m^2 \ldots r_{\prime}^2}.$$

It is to be noted, both in these and the foregoing expressions, that $p_0 = p$; $u_0 = u.$

(VII.) The expression for u_m may be used to simplify the co-ordinates of the mth point of contact; for, since

$$\frac{1}{u_m^4\, u_{m-1}^4 \ldots u^4} = \Sigma \left(\frac{x^2}{a^{4m+6}}\right),$$

let the suffix m be lowered by unity; then, assuming v to be a line such that

$$\frac{1}{v^{4m}} = \Sigma \left(\frac{x^2}{a^{4m+2}}\right)$$

we have, $u_{m-1}^2\, u_{m-2}^2 \ldots u^2 = v^{2m}.$

Hence the values of $x_m\, y_m\, z_m$ in (IV) become,

$$x_m = \left(\frac{v}{a}\right)^{2m} x.$$

$$y_m = \left(\frac{v}{b}\right)^{2m} y.$$

$$z_m = \left(\frac{v}{c}\right)^{2m} z.$$

(VIII.) A plane A B C (fig. 1), being drawn through corresponding extreme points of the three axes of the ellipsoid, we proceed to ascertain the co-ordinates of the points in which it is pierced by consecutive tangent-perpendiculars $p\ p_{,}\ \dots\ p_{m}$, generated in the manner before described. It will be seen that this investigation leads to a series of equations similar in form to those of (VI), but containing only the *uneven* powers of $a\ b\ c$ in the denominators; those which have already been determined containing only the *even* powers.

If $\xi\ \eta\ \zeta$ are co-ordinates of the plane, it may be represented by the equation,

$$\frac{\xi}{a} + \frac{\eta}{b} + \frac{\zeta}{c} = 1 \qquad (1)$$

The radius vector r to the first point $x y z$ will be given by the two equations,

$$z\xi - x\zeta = 0. \qquad z\eta - y\zeta = 0. \qquad (2)$$

At the point in which the given plane is pierced by this line $\xi\ \eta\ \zeta$ are coincident in (1) and (2), and we obtain,

$$\left(\frac{x}{a} + \frac{y}{b} + \frac{z}{c}\right)\frac{\zeta}{z} = 1.$$

Now assume w a number, such that,

$$\frac{x}{a} + \frac{y}{b} + \frac{z}{c} = w;$$

then, for the intersection of r with the plane under consideration,

$$\xi = \frac{x}{w}. \qquad \eta = \frac{y}{w}. \qquad \zeta = \frac{z}{w}.$$

Again, if $x_{,}\ y_{,}\ z_{,}$, as before, are the co-ordinates of the point in which the first tangent-perpendicular p pierces the ellipsoid, $\xi_{,}\ \eta_{,}\ \zeta_{,}$ those of the point in which it meets this plane; we shall have,

$$\xi_{,} = \frac{x_{,}}{w_{,}}; \qquad \eta_{,} = \frac{y_{,}}{w_{,}}; \qquad \zeta_{,} = \frac{z_{,}}{w};$$

$$w_{,} = \frac{x_{,}}{a} + \frac{y_{,}}{b} + \frac{z_{,}}{c};$$

c

and, generally, for the mth tangent-perpendicular p_m,

$$\xi_m = \frac{x_m}{w_m}; \quad \eta_m = \frac{y_m}{w_m}; \quad \zeta_m = \frac{z_m}{w_m};$$

$$w_m = \frac{x_m}{a} + \frac{y_m}{b} + \frac{z_m}{c};$$

in which it is to be noted that $w_0 = w$, &c. Hence we obtain, using the forms of (VI),

$$\xi_m = \left(\frac{v}{a}\right)^{2m}\frac{x}{w_m}; \quad \eta_m = \left(\frac{v}{b}\right)^{2m}\frac{y}{w_m}; \quad \zeta_m = \left(\frac{v}{c}\right)^{2m}\frac{z}{w_m};$$

and,

$$w_m = v^{2m}\ \Sigma\left(\frac{x}{a^{2m+1}}\right):$$

$$\therefore\ \xi_m = \frac{x}{a^{2m}\Sigma\left(\frac{x}{a^{2m+1}}\right)}. \quad \eta_m = \frac{y}{b^{2m}\Sigma\left(\frac{x}{a^{2m+1}}\right)}. \quad \zeta_m = \frac{z}{c^{2m}\Sigma\left(\frac{x}{a^{2m+1}}\right)}.$$

Now, let ρ_m be the radius vector of that point in which the plane is intersected by the mth perpendicular, so that ρ_1 is measured on p, and ρ_0, or ρ, on r; then

$$\rho_m^2 = \xi_m^2 + \eta_m^2 + \zeta_m^2,$$

$$\therefore\ \rho_m^2\ \Sigma^2\left(\frac{x}{a^{2m+1}}\right) = \Sigma\left(\frac{x^2}{a^{4m}}\right). \tag{3}$$

It has been shewn that, for the ellipsoid, (VI), page 8,

$$\Sigma\left(\frac{x^2}{a^{4m+4}}\right) = \frac{1}{p_m^2\ p_{m-1}^2\ \cdots\ p^2\ r_m^2\ r_{m-1}^2\ \cdots\ r_1^2};$$

therefore, after lowering the index m in this expression by unity, substituting in the equation (3), and extracting the square root, we find the formula, involving only the *uneven* powers of $a\ b\ c$,

$$\Sigma\left(\frac{x}{a^{2m+1}}\right) = \frac{1}{p_m\ p_{m-1}\ p_{m-2}\cdots p\ r_{m-1}\ r_{m-2}\cdots r_1}. \tag{4}$$

Cor. If this expression is squared,

$$\Sigma\left(\frac{x^2}{a^{4m+2}}\right) + 2\Sigma\left(\frac{xy}{a^{2m+1}\,b^{2m+1}}\right) = \frac{1}{\rho_m{}^2\,\rho_m{}^2{}_{-1}\cdots p^2\,r_m{}^2{}_{-1}\cdots r_l{}^2};$$

but from the second of the general formulæ in (VI) after lowering m by unity,

$$\Sigma\left(\frac{x^2}{a^{4m+2}}\right) = \frac{1}{\rho_m{}^2{}_{-1}\cdots p^2\,r_m{}^2\,r_m{}^2{}_{-1}\cdots r_l{}^2},$$

$$\therefore \quad 2\Sigma\left(\frac{xy}{a^{2m+1}\,b^{2m+1}}\right) = \frac{r_m{}^2 - \rho_m{}^2}{\rho_m{}^2\,\rho_m{}^2{}_{-1}\cdots p^2\,r_m{}^2\,r_m{}^2{}_{-1}\cdots r_l{}^2}. \quad (5)$$

(IX.) The expressions (4) and (5) in (VIII), taken together with the general forms of (VI), contain the relations which it has been our object to demonstrate as connecting these singular lines in the ellipsoid. They will be afterwards reproduced, as particular cases of the formulæ which belong to the general surface; and, in the mean time, we shall close this chapter with some remarks which suggest themselves as corollaries to the results which have been already determined.

(1.) If the axes are assumed to be equal, the surface is a sphere, and $p_{m-1}\cdots p\,r_{m-1}\cdots r_l = a^{2m-1}$; the form (4) in (VIII) will then give,

$$\frac{\Sigma(x)}{a^{2m+1}} = \frac{1}{\rho_m\,a^{2m-1}}$$

$$\therefore \quad \rho_m = \frac{a^2}{x+y+z}. \quad (1)$$

which can be readily verified.

(2.) If v_l is a line, such that,

$$v_l{}^{2m} = \rho_m\,p_{m-1}\cdots p\,r_{m-1}\cdots r_l,$$

the general co-ordinates of (VIII) may be written

$$\xi_m = \left(\frac{v_l}{a}\right)^{2m} x\,; \quad \eta_m = \left(\frac{v_l}{b}\right)^{2m} y\,; \quad \zeta_m = \left(\frac{v_l}{c}\right)^{2m} z\,;$$

but in (VII) we have deduced expressions, precisely analogous to these, for x_m y_m z_m, in which

$$v^{2m} \ = \ p_{m-1} \ \ldots \ p \ r_m \ \ldots \ r_{,} :$$

consequently,

$$\left(\frac{v_{,}}{v}\right)^{2m} \ = \ \frac{\rho_m}{r_m} : \tag{2}$$

and,

$$\frac{\xi_m}{x_m} \ = \ \frac{\eta_m}{y_m} \ = \ \frac{\zeta_m}{z_m} \ = \ \frac{\rho_m}{r_m}. \tag{3}$$

(3) By combining the general formulæ, so as to determine the absolute values of the radii and perpendiculars involved in them, in terms of the first co-ordinates, we obtain,

$$r_m^2 \ = \ \frac{\Sigma\left(\dfrac{x^2}{a^{4m}}\right)}{\Sigma\left(\dfrac{x^2}{a^{4m+2}}\right)}.$$

$$p_m^2 \ = \ \frac{\Sigma\left(\dfrac{x^2}{a^{4m+2}}\right)}{\Sigma\left(\dfrac{x^2}{a^{4m+4}}\right)}.$$

$$\rho_m^2 \ = \ \frac{\Sigma\left(\dfrac{x^2}{a^{4m}}\right)}{\Sigma^2\left(\dfrac{x}{a^{2m+1}}\right)}.$$

Hence we have, further,

$$r_m^2 \, p_m^2 \ = \ \frac{\Sigma\left(\dfrac{x^2}{a^{4m}}\right)}{\Sigma\left(\dfrac{x^2}{a^{4m+4}}\right)}.$$

$$\frac{\rho_m^2}{r_m^2} \ = \ \frac{\Sigma\left(\dfrac{x}{a^{2m+1}}\right)^2}{\Sigma^2\left(\dfrac{x}{a^{2m+1}}\right)}.$$

Moreover, since $u_m = p_m\, r_{m+1}$, if the value of m in r_m is increased by unity, there will be found,

$$u_m{}^4 = \frac{\Sigma\left(\dfrac{x^2}{a^{4m+2}}\right)}{\Sigma\left(\dfrac{x^2}{a^{4m+6}}\right)}.$$

(4.) In the equations of (VI), if all the lines are referred to a numerical unit, we have the following summation to m terms;

$$1 + \frac{1}{p^2} + \frac{1}{p^2\,r_{,}{}^2} + \frac{1}{p^2 p_{,}{}^2 r_{,}{}^2} + \frac{1}{p^2 p_{,}{}^2 r_{,}{}^2 r_{,}{}^2} + \dots$$

$$= \Sigma\left\{\frac{x^2}{a^2} + \frac{x^2}{a^4} + \frac{x^2}{a^6} + \frac{x^2}{a^8} + \dots\right\}$$

$$= \frac{a^{2m}-1}{a^2-1}\,\frac{x^2}{a^{2m}} + \frac{b^{2m}-1}{b^2-1}\,\frac{y^2}{b^{2m}} + \frac{c^{2m}-1}{c^2-1}\,\frac{z^2}{c^{2m}}. \qquad (4)$$

If m is supposed to become infinite,

$$1 + \frac{1}{p^2} + \frac{1}{p^2\,r_{,}{}^2} + \frac{1}{p^2 p_{,}{}^2 r_{,}{}^2} + \dots \ ad\ \infty,$$

$$= \frac{x^2}{a^2-1} + \frac{y^2}{b^2-1} + \frac{z^2}{c^2-1}. \qquad (5)$$

Again, the developement of (4) in (VIII) will give the series, to m terms,

$$\frac{r}{\rho} + \frac{1}{\rho_{,}p} + \frac{1}{\rho_2\, p_{,} p r_{,}} + \frac{1}{\rho_3\ p_2\ p_{,}\ p\ r_2 r_{,}} + \dots$$

$$= \Sigma\left\{\frac{x}{a} + \frac{x}{a^3} + \frac{x}{a^5} + \frac{x}{a^7} + \dots\right\}$$

$$= \frac{a^{2m}-1}{a^2-1}\,\frac{x}{a^{2m-1}} + \frac{b^{2m}-1}{b^2-1}\,\frac{y}{b^{2m-1}} + \frac{c^{2m}-1}{c^2-1}\,\frac{z}{c^{2m-1}}; \qquad (6)$$

and, if $m = \infty$,

$$\frac{r}{\rho} + \frac{1}{\rho_{,}p} + \frac{1}{\rho_2 p_{,} p r_{,}} + \dots \ ad\ \infty,$$

$$= \frac{ax}{a^2-1} + \frac{by}{b^2-1} + \frac{cz}{c^2-1}. \qquad (7)$$

CHAPTER II.

(I.) We proceed to extend this investigation to the case of the general class of surfaces represented by the equation,

$$\frac{x^n}{a^n} + \frac{y^n}{b^n} + \frac{z^n}{c^n} = 1. \tag{1}$$

The tangent-plane at any point $x\,y\,z$ is given by,

$$\frac{x^{n-1}}{a^n}\xi + \frac{y^{n-1}}{b^n}\eta + \frac{z^{n-1}}{c^n}\zeta = 1, \tag{2}$$

$\xi\,\eta\,\zeta$ being the general co-ordinates of the plane.

The perpendicular upon this plane from the origin is defined by the equations,

$$a^n z^{n-1}\xi - c^n x^{n-1}\zeta = 0; \quad b^n z^{n-1}\eta - c^n y^{n-1}\zeta = 0; \tag{3}$$

$\xi\,\eta\,\zeta$ being the general co-ordinates of the line.

For the point in which the surface is intersected by the perpendicular we write the co-ordinates x_i, y_i, z_i, and there are, for the determination of these quantities, the three conditions,

$$\frac{x_i^n}{a^n} + \frac{y_i^n}{b^n} + \frac{z_i^n}{c^n} = 1.$$

$$a^n z^{n-1} x_i - c^n x^{n-1} z_i = 0. \quad b^n z^{n-1} y_i - c^n y^{n-1} z_i = 0.$$

Assume u to be a line, such that

$$\frac{1}{u^{2n}} = \frac{x^{n(n-1)}}{a^{n(n+1)}} + \frac{y^{n(n-1)}}{b^{n(n+1)}} + \frac{z^{n(n-1)}}{c^{n(n+1)}}; \tag{4}$$

then the elimination will give,

$$\left(\frac{z_{,}}{u^2}\right)^n = \left(\frac{z^{n-1}}{c^n}\right)^n :$$

from which we obtain, taking the positive sign,

$$
\left.
\begin{aligned}
z_{,} &= \frac{u^2 z^{n-1}}{c^n} \\[2mm]
y_{,} &= \frac{u^2 y^{n-1}}{b^n} \\[2mm]
x_{,} &= \frac{u^2 x^{n-1}}{a^n}
\end{aligned}
\right\}
\qquad (5)
$$

Again, if $x'\, y'\, z'$ are the co-ordinates of intersection between the tangent plane and its perpendicular,

$$\frac{x^{n-1}}{a^n} x' + \frac{y^{n-1}}{b^n} y' + \frac{z^{n-1}}{c^n} z' = 1.$$

$$a^n z^{n-1} x' - c^n x^{n-1} z' = 0. \qquad b^n z^{n-1} y' - c^n y^{n-1} z' = 0.$$

Assume v to be a line, such that,

$$\frac{1}{v^2} = \frac{x^{2(n-1)}}{a^{2n}} + \frac{y^{2(n-1)}}{b^{2n}} + \frac{z^{2(n-1)}}{c^{2n}} ;$$

we shall then obtain for the intersection,

$$
\left.
\begin{aligned}
z' &= \frac{v^2 z^{n-1}}{c^n} \\[2mm]
y' &= \frac{v^2 y^{n-1}}{b^n} \\[2mm]
x' &= \frac{v^2 x^{n-1}}{a^n}
\end{aligned}
\right\}
\qquad (6)
$$

Now, let p be the perpendicular upon the tangent plane, r, the intercept on p between the origin and the surface ; i.e. the radius vector of the point of intersection : then,

$$p^2 = x'^2 + y'^2 + z'^2$$

$$= v^4 \left\{ \frac{x^{2(n-1)}}{a^{2n}} + \frac{y^{2(n-1)}}{b^{2n}} + \frac{z^{2(n-1)}}{c^{2n}} \right\} = v^2 :$$

$$\therefore \quad v \;=\; p$$

and,
$$\frac{1}{p^2} \;=\; \Sigma \left\{ \frac{x^{2(n-1)}}{a^{2n}} \right\}.$$

Also,
$$r_i^2 \;=\; x_i^2 \;+\; y_i^2 \;+\; z_i^2,$$
$$\;=\; \frac{u^4}{p^2},$$
$$\therefore \quad u^2 \;=\; p r_i \,;$$

the line u is therefore a mean proportional between the perpendicular and its intercept, as in the ellipsoid.

(II.) From what precedes we have the following simultaneous equations, involving $r\,p\,r_i$;

$$\frac{x^n}{a^n} \;+\; \frac{y^n}{b^n} \;+\; \frac{z^n}{c^n} \;=\; 1.$$

$$\frac{x^{2(n-1)}}{a^{2n}} \;+\; \frac{y^{2(n-1)}}{b^{2n}} \;+\; \frac{z^{2(n-1)}}{c^{2n}} \;=\; \frac{1}{p^2}.$$

$$\frac{x^{n(n-1)}}{a^{n(n+1)}} \;+\; \frac{y^{n(n-1)}}{b^{n(n+1)}} \;+\; \frac{z^{n(n-1)}}{c^{n(n+1)}} \;=\; \frac{1}{p^n r_i^n}.$$

$$x^2 \;+\; y^2 \;+\; z^2 \;=\; r^2.$$

The elimination of $x\,y\,z$ in the general case does not appear to be practicable; in the case of the ellipsoid, where $n = 2$, it has already been effected, and, in any other individual instance, will lead to the corresponding relation $f(r\,p\,r_i) = 0$.

(III.) Let us now consider a series of points to be determined according to the same law which was before supposed to obtain in the ellipsoid, that is by the intersection of the surface with perpendiculars upon successive tangent planes; and let $x_m y_m z_m$ be co-ordinates of the mth point of intersection, r_m its radius vector, $p_m\, u_m$ the corresponding symbols for p and u: there is, then, by generalising the formulæ of (I), (5); Chapter II :

$$x_m \;=\; \frac{u_{m-1}^2\, x_{m-1}^{n-1}}{a^n}. \qquad y_m \;=\; \frac{u_{m-1}^2\, y_{m-1}^{n-1}}{b^n}. \qquad z_m \;=\; \frac{u_{m-1}^2\, z_{m-1}^{n-1}}{c^n}.$$

As the quantities to be determined are all identical in form, it will be sufficient to consider one of them only, as x_m, and we find,

$$x_m = \frac{u_{m-1}^2 \; u_{m-2}^{2(n-1)} \; u_{m-3}^{2(n-1)^2} \; \ldots \; u_{,}^{2(n-1)^{m-2}} \; u^{2(n-1)^{m-1}}}{a^n \; a^{n(n-1)} \; a^{n(n-1)^2} \ldots \; a^{n(n-1)^{m-2}} \; a^{n(n-1)^{m-1}}} \; x^{(n-1)^m}.$$

The number of factors in each term of this fraction is m, and the indices of the denominator form the geometric progression,

$$n \;+\; n(n-1) \;+\; n(n-1)^2 \;+\; \ldots \;=\; n\,\frac{(n-1)^m-1}{n-2}$$

$$\therefore \quad x_m \;=\; \frac{u_{m-1}^2 \; u_{m-2}^{2(n-1)} \ldots u_{,}^{2(n-1)^{m-2}} \; u^{2(n-1)^{m-1}} \; x^{(n-1)^m}}{a^{\,n\frac{(n-1)^m-1}{n-2}}} \tag{1}$$

with similar expressions for $y_m \; z_m$.

For determining the consecutive perpendiculars there are the relations,

$$\frac{1}{p^2} \;=\; \Sigma \left\{ \frac{x^{2(n-1)}}{a^{2n}} \right\}; \quad \frac{1}{p_{,}^2} \;=\; \Sigma \left\{ \frac{x_{,}^{2(n-1)}}{a^{2n}} \right\}; \quad \ldots$$

and, in general terms,

$$\frac{1}{p_m^2} \;=\; \Sigma \left\{ \frac{x_m^{2(n-1)}}{a^{2n}} \right\}. \tag{2}$$

Similarly, for determining the lines $u \; u_{,} \ldots u_m$ we have, in general,

$$\frac{1}{u_m^{2n}} \;=\; \Sigma \left\{ \frac{x_m^{n(n-1)}}{a^{n(n+1)}} \right\}. \tag{3}$$

D

Now, if the values before ascertained are assigned to the co-ordinates,

$$\frac{1}{p_m{}^2} = \Sigma \left\{ \frac{4(n-1) \; 4(n-1)^2 \; \ldots \; 4(n-1)^m \; 2(n-1)^{m+1}}{u_{m-1} \quad u_{m-2} \; \ldots \; u \quad x} \right. \left. \frac{}{a^{2n} \quad a^{2n(n-1)} \ldots \; a^{2n(n-1)^{m-1}} \; a^{2n(n-1)^m}} \right\} . \tag{4}$$

$$\frac{1}{u_m{}^{2n}} = \Sigma \left\{ \frac{2n(n-1) \; 2n(n-1)^2 \; \ldots \; 2n(n-1)^m \; n(n-1)^{m+1}}{u_{m-1} \quad u_{m-2} \; \ldots \; u \quad x} \right. \left. \frac{}{a^{n^2(n-1)} \; a^{n^2(n-1)^2} \; \ldots \; a^{n^2(n-1)^m} \; a^{n(n+1)}} \right\} . \tag{5}$$

In the expression (4) for p_m it is to be remarked that the number of factors, both in the numerator and denominator, is $(m + 1)$, and we have, for the index of a in the denominator, the geometric series,

$$2n + 2n (n-1) + 2n (n-1)^2 + \ldots = 2n \frac{(n-1)^{m+1} - 1}{n-2}.$$

The number of factors in each term of the expression (5) for u_m is also $(m+1)$, and we find, for the index of a, the series,

$$n^2(n-1) + n^2(n-1)^2 + \ldots \text{ to } m \text{ terms} + n(n+1) = \frac{n}{n-2} \left\{ n(n-1)^{m+1} - 2 \right\}.$$

Hence, the general formulæ become,

$$\frac{1}{p_m{}^2} = \Sigma \left\{ \frac{4(n-1) \; 4(n-1)^2 \; \ldots \; 4(n-1)^m \; 2(n-1)^{m+1}}{u_{m-1} \quad u_{m-2} \; \ldots \; u \quad x} \Bigg/ \; 2n \frac{(n-1)^{m+1} - 1}{n-2} \atop a \right\} . \tag{6}$$

$$\frac{1}{u_m{}^{2n}} = \Sigma \left\{ \frac{2n(n-1) \; 2n(n-1)^2 \; \ldots \; 2n(n-1)^m \; n(n-1)^{m+1}}{u_{m-1} \quad u_{m-2} \; \ldots \; u \quad x} \Bigg/ \; n \frac{n(n-1)^{m+1} - 2}{n-2} \atop a \right\} . \tag{7}$$

It has been already shewn, page 16, that $u^2 = p\,r_{_1}$; consequently, $u_m^2 = p_m\,r_{m+1}$; $u^2_{m-1} = p_{m-1}\,r_m$ ... : and, after reducing by means of these equivalent forms, we shall obtain the general expressions,

$$\Sigma \left\{ \frac{x^{2(n-1)}{}^{m+1}}{\underset{a}{2n\frac{(n-1)-1}{n-2}}{}^{m+1}} \right\} = \frac{1}{\underset{p_m}{2}\; \underset{p_{m-1}}{2(n-1)}\; \cdots\; \underset{p}{2(n-1)}{}^{m}\; \underset{r_m}{2(n-1)}\; \underset{r_{m-1}}{2(n-1)}{}^{2}\; \cdots\; \underset{r_{_1}}{2(n-1)}{}^{m}}.$$

$$\Sigma \left\{ \frac{x^{n(n-1)}{}^{m+1}}{\underset{a}{n\frac{n(n-1)-2}{n-2}}{}^{m+1}} \right\} = \frac{1}{\underset{p_m}{n}\; \underset{p_{m-1}}{n(n-1)}\; \cdots\; \underset{p}{n(n-1)}{}^{m}\; \underset{r_{m+1}}{n}\; \underset{r_m}{n(n-1)}\; \cdots\; \underset{r_{_1}}{n(n-1)}{}^{m}}.$$

(IV.) We have now to shew that, for the general surface,

$$x_m = \left\{ \frac{v^2}{a^n} \right\}^{\frac{(n-1)^m - 1}{n-2}} x^{(n-1)^m};$$

with similar expressions for y_m z_m: v being a line such that,

$$\frac{1}{v^{\frac{2n\frac{(n-1)^m - 1}{n-2}}{}}} = \Sigma \left\{ \frac{x^{n(n-1)^m}}{\underset{a}{n\frac{n(n-1)-2}{n-2}}} \right\}. \tag{1}$$

From page 17, (1),

$$x_m = \frac{u_{m-1}^{2}\; u_{m-2}^{2(n-1)}\; \cdots\; u^{2(n-1)^{m-1}}\; x^{(n-1)^m}}{\underset{a}{n\frac{(n-1)^m - 1}{n-2}}}.$$

Again, from page 18, (7),

$$\frac{1}{u_m{}^{2n}} \;=\; \Sigma\left\{ \frac{u_{m-1}^{2n(n-1)}\, u_{m-2}^{2n(n-1)^2} \dots u^{2n(n-1)^m}\, x^{u(n-1)^{m+1}}}{a^{\,n\frac{n(n-1)^{m+1}-2}{n-2}}} \right\}.$$

Now, after extracting the nth root of this expression, the index m being lowered by unity, we obtain,

$$\frac{1}{u_{m-1}^{2}\, u_{m-2}^{2(n-1)} \dots u_{,}^{2(n-1)^{m-2}}\, u^{2(n-1)^{m-1}}} \;=\; \Sigma^{\frac{1}{n}}\left\{ \frac{x^{n(n-1)^m}}{a^{\,n\frac{n(n-1)^m-2}{n-2}}} \right\};$$

$$=\; \frac{1}{v^{\,2\frac{(n-1)^m-1}{n-2}}};$$

$$\therefore\; u_{m-1}^{2}\, u_{m-2}^{2(n-1)} \dots u^{2(n-1)^{m-1}} \;=\; v^{\,2\frac{(n-1)^m-1}{n-2}}. \tag{2}$$

Hence there will be obtained the following values of the co-ordinates of the mth point on the surface, expressed in terms of those of the first assumed point,

$$x_m \;=\; \left\{\frac{v^2}{a^n}\right\}^{\frac{(n-1)^m-1}{n-2}} x^{(n-1)^m}.$$

$$y_m \;=\; \left\{\frac{v^2}{b^n}\right\}^{\frac{(n-1)^m-1}{n-2}} y^{(n-1)^m}. \tag{3}$$

$$z_m \;=\; \left\{\frac{v^2}{c^n}\right\}^{\frac{(n-1)^m-1}{n-2}} z^{(n-1)^m}.$$

Cor. If $m = 0$ these expressions will become,

$$x_0 = x \; ; \quad y_0 = y \; ; \quad z_0 = z \; ;$$

as they ought to do consistently with the previous assumptions; while the formula for v, (1), will be resolved into the equation of the surface,

$$\Sigma \left(\frac{x}{a} \right)^n = 1.$$

(V.) It remains, further, to investigate the formula for the general surface, involving ρ_m, which corresponds to (4) page 10, derived from the ellipsoid.

The co-ordinates of the mth point of intersection with the plane $\Sigma \left(\frac{x}{a} \right) = 1$ will be, as before,

$$\xi_m = \frac{x_m}{w_m} \; ; \quad \eta_m = \frac{y_m}{w_m} \; ; \quad \zeta_m = \frac{z_m}{w_m} \; ;$$

$$w_m = \Sigma \left(\frac{x_m}{a} \right).$$

Now, employing the results of the preceding section (IV),

$$x_m = \frac{v^{2^{\frac{(u-1)-1}{n-2}m}}}{a \, n^{\frac{(n-1)-1}{n-2}m}} \; x^{(n-1)^m} \; ; \quad \&c.$$

$$w_m = \frac{v^{2^{\frac{(n-1)-1}{n-2}m}}}{a \, n^{\frac{(n-1)-1}{n-2}m}} \; \frac{x^{(n-1)^m}}{a} \; + \; \cdots$$

$$\therefore \quad w_m = v^{2\frac{(n-1)-1}{n-2}^m} \Sigma \left\{ \frac{x^{(n-1)^m}}{a^{\frac{n(n-1)-2}{n-2}^m}} \right\} : \tag{1}$$

$$\therefore \quad \xi_m = \frac{x^{(n-1)^m}}{a^{n\frac{(n-1)-1}{n-2}^m}} \Sigma^{-1} \left\{ \frac{x^{(n-1)^m}}{a^{\frac{n(n-1)-2}{n-2}^m}} \right\} : \tag{2}$$

with similar expressions for η_m ζ_m.

But, $\quad \rho_m^2 = \xi_m^2 + \eta_m^2 + \zeta_m^2$,

$$\therefore \quad \rho_m^2 = \Sigma^{-2} \left\{ \frac{x^{(n-1)^m}}{a^{\frac{n(n-1)-2}{n-2}^m}} \right\} \Sigma \left\{ \frac{x^{2(n-1)^m}}{a^{2n\frac{(n-1)-1}{n-2}^m}} \right\}. \tag{3}$$

In order to reduce this expression, we have from page 19, after lowering the suffix m by unity,

$$\Sigma \left\{ \frac{x^{2(n-1)^m}}{a^{2n\frac{(n-1)-1}{n-2}^m}} \right\} = \frac{1}{2 \atop p_{m-1} \ldots p} \frac{1}{2(n-1)^{m-1} \atop r_{m-1} \ldots r_{,}}$$

and the substitution of this value in the preceding equation will give, after extracting the square root, the formula,

$$\Sigma \left\{ \frac{x^{(n-1)^m}}{a^{\frac{n(n-1)-2}{n-2}^m}} \right\} = \frac{1}{p_m^{(n-1)} p_{m-1}^{(n-1)} p_{m-2}^{(n-1)} \ldots p} \frac{1}{r_{m-1}^{(n-1)^{m-1}} \ldots r_{,}}. \tag{4}$$

(VI.) It will be convenient in this place to recapitulate the principal formulæ of relation which have been investigated.

We have, then, from the general surface,

$$\Sigma \left\{ \frac{x^n}{a^n} \right\} = 1.$$

From page 22, form (4), and from page 19,

$$\Sigma \left\{ \frac{x^{(n-1)^m}}{a^{\frac{n(n-1)-2}{n-2}}} \right\} = \frac{1}{p_m \, p_{m-1} \, p_{m-2} \dots p^{\frac{(n-1)}{m-1}} \quad r_{m-1}^{(n-1)} \dots r_{\prime}^{\frac{(n-1)}{m-1}}}. \quad (1)$$

$$\Sigma \left\{ \frac{x^{(n-1)^m}}{a^{\frac{n(n-1)-1}{n-2}}} \right\}^2 = \frac{1}{p_{m-1}^{2(n-1)} \, p_{m-2}^{2(n-1)} \dots p^{2} \quad r_{m-1}^{\frac{2(n-1)}{m-1}} \dots r_{\prime}^{\frac{2(n-1)}{m-1}}}. \quad (2)$$

$$\Sigma \left\{ \frac{x^{(n-1)^m}}{a^{\frac{n(n-1)-2}{n-2}}} \right\}^n = \frac{1}{p_{m-1}^{n(n-1)} \, p_{m-2}^{n(n-1)} \dots p^{n} \quad r_{m}^{n} \dots r_{\prime}^{\frac{n(n-1)}{m-1}}}. \quad (3)$$

By a combination of (1) (3) there is derived the relation,

$$\Sigma^n \left\{ \frac{x^{(n-1)^m}}{a^{\frac{n(n-1)-2}{n-2}}} \right\} - \Sigma \left\{ \frac{x^{(n-1)^m}}{a^{\frac{n(n-1)-2}{n-2}}} \right\}^n =$$

$$\frac{r_m^{n} - \rho_m^{n}}{p_m^{n} \, p_{m-1}^{n} \, p_{m-2}^{n(n-1)} \dots p^{n(n-1)} \quad r_m^{n} \, r_{m-1}^{n(n-1)} \dots r_{\prime}^{\frac{n(n-1)}{m-1}}}. \quad (4)$$

(VII.) By combining the forms of (VI) we obtain the absolute values of r_m p_m ρ_m in terms of the first co-ordinates.

Increasing m by unity in (2), and eliminating by means of (3),

$$p_m^{2n} = \dfrac{\Sigma^2 \left\{ \dfrac{x^{(n-1)^m}}{a^{\frac{n(n-1)-2}{n-2}}} \right\}^n}{\Sigma^n \left\{ \dfrac{x^{(n-1)^{m+1}}}{a^{n\frac{(n-1)-1}{n-2}}} \right\}^2}. \tag{5}$$

By (3) and (2),

$$r_m^{2n} = \dfrac{\Sigma^n \left\{ \dfrac{x^{(n-1)^m}}{a^{n\frac{(n-1)-1}{n-2}}} \right\}^2}{\Sigma^2 \left\{ \dfrac{x^{(n-1)^m}}{a^{\frac{n(n-1)-2}{n-2}}} \right\}^n}. \tag{6}$$

e

By (1) and (2),

$$\rho_m^{2} = \dfrac{\Sigma \left\{ \dfrac{x^{(n-1)^m}}{a^{n\frac{(n-1)-1}{n-2}}} \right\}^2}{\Sigma^2 \left\{ \dfrac{x^{(n-1)^m}}{a^{\frac{n(n-1)-2}{n-2}}} \right\}}. \tag{7}$$

The combination of (5) (6) will give the expression,

$$ r_m^2\, p_m^2 \;=\; \frac{\Sigma\left\{\dfrac{x^{(n-1)^m}}{a^{n\frac{(n-1)^m-1}{n-2}}}\right\}^2}{\Sigma\left\{\dfrac{x^{(n-1)^{m+1}}}{a^{n\frac{(n-1)^{m+1}-1}{n-2}}}\right\}^2}. \qquad (8)$$

By augmenting, in (5), the value of m by unity, and using the relation $u_m^2 = r_{m+1}\, p_m$,

$$ u_m^{2n} \;=\; \frac{\Sigma\left\{\dfrac{x^{(n-1)^m}}{a^{\frac{n(n-1)^m-2}{n-2}}}\right\}^n}{\Sigma\left\{\dfrac{x^{(n-1)^{m+1}}}{a^{\frac{n(n-1)^{m+1}-2}{n-2}}}\right\}^n}. \qquad (9)$$

Lastly, by combining (6) (7),

$$ \frac{p_m^n}{r_m^n} \;=\; \frac{\Sigma\left\{\dfrac{x^{(n-1)^m}}{a^{\frac{n(n-1)^m-2}{n-2}}}\right\}^n}{\Sigma^n\left\{\dfrac{x^{(n-1)^m}}{a^{\frac{n(n-1)^m-2}{n-2}}}\right\}}. \qquad (10)$$

$$ \therefore\; p_m^n\, \Sigma\left\{\dfrac{x^{(n-1)^m}}{a^{\frac{n(n-1)^m-2}{n-2}}}\right\}^n \;=\; r_m^n\, \Sigma^n\left\{\dfrac{x^{(n-1)^m}}{a^{\frac{n(n-1)^m-2}{n-2}}}\right\}. \qquad (11)$$

E

(VIII.) In adapting these general formulæ to the particular case of the Ellipsoid, the indices are found to take the singular form $\frac{0}{0}$, and their special values must be ascertained in the usual manner. Then,

$$\frac{n(n-1)\overset{m}{-}2}{n-2} \;=\; \frac{0}{0} \;=\; (n-1)^{m} \;+\; mn\,(n-1)^{m-1} \;=\; 2m+1.$$

$$n\frac{(n-1)\overset{m}{-}1}{n-2} \;=\; \frac{0}{0} \;=\; (n-1)^{m} \;+\; mn\,(n-1)^{m-1}-1 \;=\; 2m.$$

Hence the formulæ will be for the ellipsoid, as shewn independently at pages 8 and 10,

$$\Sigma\left\{\frac{x}{a^{2m+1}}\right\} \;=\; \frac{1}{\rho_m\;p_{m-1}\;p_{m-2}\;\ldots\;p\;r_{m-1}\;r_{m-2}\;\ldots\;r_{\prime}}. \tag{1}$$

$$\Sigma\left\{\frac{x^2}{a^{4m}}\right\} \;=\; \frac{1}{p_{m-1}^2\;p_{m-2}^2\;\ldots\;p^2\;r_{m-1}^2\;r_{m-2}^2\;\ldots\;r_{\prime}^2}. \tag{2}$$

$$\Sigma\left\{\frac{x^2}{a^{4m+2}}\right\} \;=\; \frac{1}{p_{m-1}^2\;p_{m-2}^2\;\ldots\;p^2\;r_m^2\;r_{m-1}^2\;\ldots\;r_{\prime}^2}. \tag{3}$$

$$2\Sigma\left\{\frac{xy}{a^{2m+1}\,b^{2m+1}}\right\} \;=\; \frac{r_m^2\;-\;\rho_m^2}{p_m^2\;p_{m-1}^2\;p_{m-2}^2\;\ldots\;p^2\;r_m^2\;r_{m-1}^2\;\ldots\;r_{\prime}^2}. \tag{4}$$

The remaining formulæ (5) — (10), for $r_m\,p_m\,\rho_m\,u_m$, will be found to agree with those which have been established in page 12.

(IX.) In this section we propose to determine the equation of the plane containing the lines $r\ p\ r_{,}$; and, generally, the equation of the plane which contains the lines $r_{m-1}\ p_{m-1}\ r_m$, in the surface of the nth degree.

Let X Y Z be the co-ordinates of the plane in which are situated the lines $r\ p\ r_{,}$; $x\ y\ z$ the co-ordinates of contact; $x_{,}\ y_{,}\ z_{,}$ those of intersection : then, since the plane passes through the origin, its equation may be written,

$$Z \quad = \quad M X \quad + \quad N Y. \tag{1}$$

Consequently, $$z \quad = \quad M x \quad + \quad N y. \tag{2}$$

$$z_{,} \quad = \quad M x_{,} \quad + \quad N y_{,}. \tag{3}$$

But, from page 15, (5),

$$x_{,} = \frac{n^2\,x^{n-1}}{a^n}; \qquad y_{,} = \frac{n^2\,y^{n-1}}{b^n}; \qquad z_{,} = \frac{n^2\,z^{n-1}}{c^n};$$

$$\therefore \frac{z^{n-1}}{c^n} = M\frac{x^{n-1}}{a^n} + N\frac{y^{n-1}}{b^n}. \tag{4}$$

By combining (4) and (2) we find.

$$M = - \frac{a^n(b^n\,z^{n-2} - c^n\,y^{n-2})}{c^n(a^n\,y^{n-2} - b^n\,x^{n-2})}\,\frac{z}{x}.$$

$$N = + \frac{b^n(a^n\,z^{n-2} - c^n\,x^{n-2})}{c^n(a^n\,y^{n-2} - b^n\,x^{n-2})}\,\frac{z}{y}.$$

The plane containing the lines $r\ p\ r_{,}$ is, therefore, represented by the equation,

$$a^n(b^n\,z^{n-2} - c^n\,y^{n-2})\,\frac{X}{x} -$$

$$b^n(a^n\,z^{n-2} - c^n\,x^{n-2})\,\frac{Y}{y} +$$

$$c^n(a^n\,y^{n-2} - b^n\,x^{n-2})\,\frac{Z}{z} = 0; \tag{5}$$

or,

$$a^n (b^n z^{n-2} - c^n y^{n-2})\, X\, y\, z\, -$$

$$b^n (a^n z^{n-2} - c^n x^{n-2})\, x\, Y\, z\, +$$

$$c^n (a^n y^{n-2} - b^n x^{n-2})\, x\, y\, Z\, =\, 0. \tag{6}$$

In the determination of the plane which contains the next three consecutive lines $r_{,}\, p_{,}\, r_2$ it is necessary to write $x_{,}\, y_{,}\, z_{,}$, as given in (5), page 15, instead of $x\, y\, z$ in either of the preceding equations, (5) or (6), there will be found after the reductions are completed, representing the second plane, the equation,

$$a^n \left\{ \frac{b^u}{c^{n(n-2)}}\, z^{(n-1)(n-2)} - \frac{c^n}{b^{n(n-2)}}\, y^{(n-1)(n-2)} \right\} \frac{a^n\, X}{x^{n-1}} -$$

$$b^u \left\{ \frac{a^n}{c^{n(n-2)}}\, z^{(n-1)(n-2)} - \frac{c^u}{a^{n(n-2)}}\, x^{(n-1)(n-2)} \right\} \frac{b^u\, Y}{y^{n-1}} +$$

$$c^n \left\{ \frac{a^u}{b^{n(n-2)}}\, y^{(n-1)(n-2)} - \frac{b^u}{a^{n(n-2)}}\, x^{(n-1)(n-2)} \right\} \frac{c^n\, Z}{z^{n-1}} = 0;$$

which is equivalent to,

$$\frac{a^{2n}}{b^{n(n-2)}\, c^{n(n-2)}} \left\{ b^{n(n-1)}\, z^{(n-1)(n-2)} - c^{n(n-1)}\, y^{(n-1)(n-2)} \right\} \frac{X}{x^{n-1}} -$$

$$\frac{b^{2n}}{a^{n(n-2)}\, c^{n(n-2)}} \left\{ a^{n(n-1)}\, z^{(n-1)(n-2)} - c^{n(n-1)}\, x^{(n-1)(n-2)} \right\} \frac{Y}{y^{n-1}} +$$

$$\frac{c^{2n}}{a^{n(n-2)}\, b^{n(n-2)}} \left\{ a^{n(n-1)}\, y^{(n-1)(n-2)} - b^{n(n-1)}\, x^{(n-1)(n-2)} \right\} \frac{Z}{z^{n-1}} = 0. \tag{7}$$

In determining the equation of the mth plane, which includes the lines r_{m-1} p_{m-1} r_m, we may substitute in (5) the general values of x_m y_m z_m given in (3), page 20; then,

$$a^n (b^n z_m^{n-2} - c^n y_m^{n-2}) \frac{X}{x_m} -$$

$$b^n (a^n z_m^{n-2} - c^n x_m^{n-2}) \frac{Y}{y_m} +$$

$$c^n (a^n y_m^{n-2} - b^n x_m^{n-2}) \frac{Z}{z_m} = 0.$$

After the reduction is completed, it will be found that the mth plane is represented by the equation,

$$a^{\frac{n}{n-2}\left\{(n-1)^m + (n-3)\right\}} \left\{ \frac{z^{(n-1)^m} (n-2)}{c^{n(n-1)}} - \frac{y^{(n-1)^m} (n-2)}{b^{n(n-1)}} \right\} \frac{b^n c^n X}{x^{(n-1)^m}} -$$

$$b^{\frac{n}{n-2}\left\{(n-1)^m + (n-3)\right\}} \left\{ \frac{z^{(n-1)^m} (n-2)}{c^{n(n-1)}} - \frac{x^{(n-1)^m} (n-2)}{a^{n(n-1)}} \right\} \frac{a^n c^n Y}{y^{(n-1)^m}} +$$

$$c^{\frac{n}{n-2}\left\{(n-1)^m + (n-3)\right\}} \left\{ \frac{y^{(n-1)^m} (n-2)}{b^{n(n-1)}} - \frac{x^{(n-1)^m} (n-2)}{a^{n(n-1)}} \right\} \frac{a^n b^n Z}{z^{(n-1)^m}} = 0. \quad (8)$$

Cor. When $n = 2$, the index $\frac{n}{n-2}\left\{(n-1)^m + (n-3)\right\} = \frac{0}{0}$; and the value of this fraction is found to be $2m$: hence we have for the plane containing the lines r_{m-1} p_{m-1} r_m in the ellipsoid, the equation,

$$a^{2m} (b^2 - c^2) \frac{X}{x} - b^{2m} (a^2 - c^2) \frac{Y}{y} + c^{2m} (a^2 - b^2) \frac{Z}{z} = 0. \quad (9)$$

(X.) If the equation (9) is developed for consecutive values of m, it will be found that the successive planes are represented by the following system of equations :

$$a^2 (b^2-c^2) \frac{X}{x} \;-\; b^2 (a^2-c^2) \frac{Y}{y} \;+\; c^2 (a^2-b^2) \frac{Z}{z} \;=\; 0.$$

$$a^4 (b^2-c^2) \frac{X}{x} \;-\; b^4 (a^2-c^2) \frac{Y}{y} \;+\; c^4 (a^2-b^2) \frac{Z}{z} \;=\; 0.$$

$$a^6 (b^2-c^2) \frac{X}{x} \;-\; b^6 (a^2-c^2) \frac{Y}{y} \;+\; c^6 (a^2-b^2) \frac{Z}{z} \;=\; 0.$$

$$a^8 (b^2-c^2) \frac{X}{x} \;-\; b^8 (a^2-c^2) \frac{Y}{y} \;+\; c^8 (a^2-b^2) \frac{Z}{z} \;=\; 0.$$

...   .

$x\,y\,z$ being the point of original contact between the ellipsoid and its first tangent plane; X Y Z the general co-ordinates in each of the planes which include the consecutive radii and perpendiculars.

(XI.) In the ellipsoid, it is required to ascertain the angle between the two planes which include consecutive systems of lines $r_{m-1}\, p_{m-1}\, r_m$; and $r_m\, p_m\, r_{m+1}$.

From equation (9) in the corollary to (IX), the equations of the consecutive planes are,

$$a^{2m} (b^2-c^2) \frac{X}{x} \;-\; b^{2m} (a^2-c^2) \frac{Y}{y} \;+\; c^{2m} (a^2-b^2) \frac{Z}{z} \;=\; 0.$$

$$a^{2m+2} (b^2-c^2) \frac{X}{x} \;-\; b^{2m+2} (a^2-c^2) \frac{Y}{y} \;+\; c^{2m+2} (a^2-b^2) \frac{Z}{z} \;=\; 0.$$

Let i be the angle of intersection between these planes, then, their equations being written,

$$\left. \begin{array}{l} Z = MX + NY \\ Z = M_{,}X + N_{,}Y \end{array} \right\},$$

we have, $\quad \cos i = \dfrac{1 + MM_{,} + NN_{,}}{(1+M^2+N^2)^{\frac{1}{2}} (1+M_{,}^2+N_{,}^2)^{\frac{1}{2}}}.$

But, $\quad M = - \dfrac{a^{2m}(b^2-c^2)}{c^{2m}(a^2-b^2)}\dfrac{z}{x};$ $\qquad N = + \dfrac{b^{2m}(a^2-c^2)}{c^{2m}(a^2-b^2)}\dfrac{z}{y};$

$$M_{,} = - \dfrac{a^{2m+2}(b^2-c^2)}{c^{2m+2}(a^2-b^2)}\dfrac{z}{x}; \quad N_{,} = + \dfrac{b^{2m+2}(a^2-c^2)}{c^{2m+2}(a^2-b^2)}\dfrac{z}{y};$$

consequently,

$$1 + MM_{,} + NN_{,} = \dfrac{z^2}{c^{4m+2}(a^2-b^2)^2} \; \Sigma \left\{ \dfrac{a^{4m+2}(b^2-c^2)^2}{x^2} \right\} :$$

$$1 + M^2 + N^2 = \dfrac{z^2}{c^{4m}(a^2-b^2)^2} \; \Sigma \left\{ \dfrac{a^{4m}(b^2-c^2)^2}{x^2} \right\} :$$

$$1 + M_{,}^2 + N_{,}^2 = \dfrac{z^2}{c^{4m+4}(a^2-b^2)^2} \; \Sigma \left\{ \dfrac{a^{4m+4}(b^2-c^2)^2}{x^2} \right\} :$$

$$\therefore (1 + M^2 + N^2)^{\frac{1}{2}} (1 + M_{,}^2 + N_{,}^2)^{\frac{1}{2}} =$$

$$\dfrac{z^2}{c^{4m+2}(a^2-b^2)^2} \; \Sigma^{\frac{1}{2}} \left\{ \dfrac{a^{4m}(b^2-c^2)^2}{x^2} \right\} \; \Sigma^{\frac{1}{2}} \left\{ \dfrac{a^{4m+4}(b^2-c^2)^2}{x^2} \right\}.$$

Hence the formula expressing the inclination will become

$$\cos i = \dfrac{\Sigma \left\{ \dfrac{a^{4m+2}(b^2-c^2)^2}{x^2} \right\}}{\Sigma^{\frac{1}{2}} \left\{ \dfrac{a^{4m}(b^2-c^2)^2}{x^2} \right\} \; \Sigma^{\frac{1}{2}} \left\{ \dfrac{a^{4m+4}(b^2-c^2)^2}{x^2} \right\}}. \qquad (10)$$

(XII.) From the equation (10) determined in (XI) it can be shewn that there exists no plane section of the ellipsoid which includes the consecutive systems of the lines $r \, p \, r_{,}$; $r_{,} \, p_{,} \, r_{,}$; ... excepting the three *principal* planes.

In the equation (10) let $\cos i = \pm 1$, a supposition implying that two consecutive sets of these lines are included in the same plane; and assume $m = 1$; then,

$$\Sigma \left\{ \frac{a^4}{x^2} (b^2 - c^2)^2 \right\} \ \Sigma \left\{ \frac{a^8}{x^2} (b^2 - c^2)^2 \right\} \ = \ \Sigma^2 \left\{ \frac{a^6}{x^2} (b^2 - c^2)^2 \right\}.$$

After effecting the reduction of this expression we shall obtain the result,

$$(a^2 - b^2)^2 \, (a^2 - c^2)^2 \, (b^2 - c^2)^2 \, (a^4 \, b^4 \, z^2 \ + \ a^4 \, c^4 \, y^2 \ + \ b^4 \, c^4 \, x^2) \ = \ 0. \quad (11)$$

If $a \, b \, c$ are unequal, as in the general case, this equation can be satisfied only by the condition,

$$\Sigma \, (a^4 \, b^4 \, z^2) \ = \ 0 \, ; \qquad\qquad (12)$$

which is true if $x = 0$, $y = 0$, $z = 0$, i.e. when the first tangent plane meets at right angles either of the three principal planes : excepting under these circumstances, therefore, the two systems of lines cannot be situated in the same plane.

When, however, there is equality between any two of the axes, i.e. if the ellipsoid is a solid of revolution, the equation (11) becomes identically true for *all* values of $x \, y \, z$; consequently in any spheroid the consecutive lines will be included always in the same plane.

Further, if the surface is changed into the hyperboloid of one or two sheets, the equation (12), which may be expressed in the form $\Sigma \left(\dfrac{a^4 \, b^4}{x^2 \, y^2} \right) = 0$, is satisfied for all points infinitely remote from the origin ; when the surfaces become identified with their conical asymptotes : and this must obviously be the case.

It is clear that the conclusion which has been here derived with respect to the first and second planes, is equally true if considered in relation to any two consecutive planes which follow them.

(XIII.) In the ellipsoid, the several tangent planes will be given by the series of equations,

$$\frac{x}{a^2}\xi + \frac{y}{b^2}\eta + \frac{z}{c^2}\zeta = 1. \tag{1}$$

$$\frac{x}{a^4}\xi + \frac{y}{b^4}\eta + \frac{z}{c^4}\zeta = \frac{1}{v^2}. \tag{2}$$

$$\frac{x}{a^6}\xi + \frac{y}{b^6}\eta + \frac{z}{c^6}\zeta = \frac{1}{v^4}. \tag{3}$$

$$\cdots \qquad \cdots \qquad \cdots$$

$$\frac{x}{a^{2m+2}}\xi + \frac{y}{b^{2m+2}}\eta + \frac{z}{c^{2m+2}}\zeta = \frac{1}{v^{2m}} \tag{m}$$

in which $v^{2m} = u^2_{m-1}\, u^2_{m-2} \ldots u^2$, as in page 8 ;

$$= p_{m-1}\, p_{m-2} \ldots p\ r_m\, r_{m-1} \ldots r_{,}.$$

In the general surface, we shall find that the mth tangent plane, which has contact at the point $x_m\, y_m\, z_m$, is represented by the equation,

$$\Sigma \left\{ \frac{x^{\frac{(n-1)}{n}\frac{m+1}{}}\,\xi}{a^{\frac{n(n-1)-1}{n-2}\frac{m+1}{}}} \right\}^2 = \frac{1}{(v^2)^{\frac{(n-1)}{n-2}-\frac{(n-1)}{}\frac{m+1}{}}} : \tag{M}$$

in which the symbol v has the same value as in (2) page 20.

If i is the angle of inclination between any two consecutive tangent planes, it may be determined from the formula (M). Or, since this inclination is the same as that of the perpendiculars upon the same planes, there is,

$$\cos i = \frac{p_m}{r_m};$$

Y

consequently, by (8) page 25,

$$\cos^2 i = \frac{\Sigma^2 \left\{ \dfrac{x^{\frac{(n-1)}{m}}}{a^{\frac{n(n-1)-2}{n-2}}} \right\}^n}{\Sigma \left\{ \dfrac{x^{\frac{(n-1)}{m+1}}}{a^{n\frac{(n-1)-1}{n-2}}} \right\}^2 \; \Sigma \left\{ \dfrac{x^{\frac{(n-1)}{m}}}{a^{n\frac{(n-1)-1}{n-2}}} \right\}^2}.$$

In the ellipsoid this expression will become,

$$\cos^2 i = \frac{\Sigma^2 \left(\dfrac{x^2}{a^{4m+2}} \right)}{\Sigma \left(\dfrac{x^2}{a^{4m+4}} \right) \; \Sigma \left(\dfrac{x^2}{a^{4m}} \right)}.$$

If $n = 2$ the successive developments of the formula (M) are easily identified with the tangent planes of the ellipsoid.

(XIV.) REMARK.—If R' and R'' are principal radii of curvature of the ellipsoid at the initial point $x\,y\,z$, they may be determined from the equation,*

$$R^2 - \frac{a^2 + b^2 + c^2 - r^2}{p} + \frac{a^2\,b^2\,c^2}{p^4} = 0 ;$$

consequently,

$$R' . R'' = \frac{a^2\,b^2\,c^2}{p^4}. \qquad R' + R'' = \frac{\Sigma\,(a^2 - x^2)}{p}.$$

Hence, if $R'\,R'' \ldots R'_m\,R''_m$ are the radii of curvature at the consecutive points of contact, there will follow,

* See Hymers' Analytical Geometry of Three Dimensions, page 265.

$$R'_m . R''_m = \frac{a^2 b^2 c^2}{p_m^4} = a^2 b^2 c^2 \frac{\Sigma^2 \left\{ \dfrac{x^2}{a^{4m+4}} \right\}}{\Sigma^2 \left\{ \dfrac{x^2}{a^{4m+2}} \right\}}. \tag{1}$$

$$R'_m + R''_m = \frac{\Sigma (a^2 - x^2)}{p_m} = \Sigma (a^2 - x^2) \frac{\Sigma^{\frac{1}{2}} \left\{ \dfrac{x^2}{a^{4m+4}} \right\}}{\Sigma^{\frac{1}{4}} \left\{ \dfrac{x^2}{a^{4m+2}} \right\}}. \tag{2}$$

By (1) and (2) the radii of curvature at the mth point of contact are expressed as functions of the co-ordinates of the initial point; and the solution will give, retaining the abbreviated form p_m, which has been determined at page 12,

$$R'_m = \frac{1}{2 p_m^2} \left\{ p_m \Sigma (a^2 - x^2) + \sqrt{p_m^2 \Sigma^2 (a^2 - x^2) - 4 a^2 b^2 c^2} \right\}.$$

$$R''_m = \frac{1}{2 p_m^2} \left\{ p_m \Sigma (a^2 - x^2) - \sqrt{p_m^2 \Sigma^2 (a^2 - x^2) - 4 a^2 b^2 c^2} \right\}.$$

(XV.) To conclude, it is proper to adapt the general forms of page 23 to those surfaces in which n is negative.

Each formula will be susceptible of two expressions, according as m is even or uneven.

1st. If m is an *even* number, all the terms which have the indices $(n-1)^{m}$, $(n-1)^{m-2}$, ... remain in the denominators; while those which involve the indices $(n-1)^{m-1}$, $(n-1)^{m-3}$ ... are transferred to the numerators : on the second side of each equation.

2nd. If m is an *uneven* number, this transference will be reversed.

Let m be an *even* number, the first three forms of page 23 will then become,

$$\Sigma \left\{ \begin{array}{c} (n+1)^{\tfrac{m}{}} \\ \dfrac{x}{} \\ n(n+1)+2 \\ a^{\tfrac{}{n+2}} \end{array} \right\} = \dfrac{(n+1)\ (n+1)^{3}\ (n+1)^{m-1}\ (n+1)^{3}\ (n+1)\ (n+1)^{m-1}}{p_{m-2}\ p_{m-4}\ \dots\ p\quad r_{m-1}\ r_{m-3}\ \dots\ r_{\prime}} \cdot \dfrac{}{(n+1)^{2}\ (n+1)^{m-2}\ (n+1)^{2}\ (n+1)^{4}\ (n+1)^{m-2}} \quad (1)$$

$$\Sigma \left\{ \begin{array}{c} (n+1)^{\tfrac{m}{}} \\ x \\ n\dfrac{(n+1)}{}-1 \\ a^{\tfrac{}{n+2}} \end{array} \right\} 2 = \dfrac{2(n+1)\ 2(n+1)^{3}\ 2(n+1)^{m-1}\ 2(n+1)\ 2(n+1)^{m-1}}{p_{m-2}\ p_{m-4}\ \dots\ p\quad r_{m-1}\ \dots\ r_{\prime}} \cdot \dfrac{}{2\quad 2(n+1)^{2}\ 2(n+1)^{m-2}\ 2(n+1)^{2}\ 2(n+1)^{m-2}} \quad (2)$$

$$\Sigma \left\{ \begin{array}{c} a^{\tfrac{n(n+1)+2}{n+2}} \\ \\ (n+1)^{m} \\ x \end{array} \right\} n = \dfrac{n\quad n(n+1)^{2}\ n(n+1)^{m-2}\ n\quad n(n+1)^{2}\ n(n+1)^{m-2}}{p_{m-1}\ p_{m-3}\ \dots\ p_{\prime}\quad r_{m}\ r_{m-2}\ \dots\ r_{2}} \cdot \dfrac{}{n(n+1)\ n(n+1)^{3}\ n(n+1)^{m-1}\ n(n+1)^{3}\ n(n+1)^{m-1}} \quad (3)$$

Let m be *uneven*,

$$\Sigma \left\{ \begin{array}{c} a^{\tfrac{n(n+1)-2}{n+2}} \\ \\ (n+1)^{m} \\ x \end{array} \right\} = \dfrac{(n+1)\ (n+1)^{3}\ (n+1)^{m-2}\ (n+1)\ (n+1)^{m-2}}{p_{m-2}\ p_{m-4}\ \dots\ p_{\prime}\quad r_{m-1}\ \dots\ r_{2}} \cdot \dfrac{}{(n+1)^{2}\ (n+1)^{m-1}\ (n+1)^{2}\ (n+1)^{m-1}} \quad (1)'$$

$$\Sigma \left\{ \begin{array}{c} a^{\tfrac{n(n+1)+1}{n+2}} \\ \\ (n+1)^{m} \\ x \end{array} \right\} 2 = \dfrac{2(n+1)\ 2(n+1)^{3}\ 2(n+1)^{m-2}\ 2(n+1)\ 2(n+1)^{m-2}}{p_{m-2}\ p_{m-4}\ \dots\ p_{\prime}\quad r_{m-1}\ \dots\ r_{2}} \cdot \dfrac{}{2\quad 2(n+1)^{2}\ 2(n+1)^{m-1}\ 2(n+1)^{2}\ 2(n+1)^{m-1}} \quad (2)'$$

$$\Sigma \left\{ \begin{array}{c} (n+1)^{\tfrac{m}{}} \\ \dfrac{x}{} \\ n(n+1)-2 \\ a^{\tfrac{}{n+2}} \end{array} \right\} n = \dfrac{n\quad n(n+1)^{2}\ n(n+1)^{m-1}\ n\quad n(n+1)^{2}\ n(n+1)^{m-1}}{p_{m-1}\ p_{m-3}\ \dots\ p\quad r_{m}\ r_{m-2}\ \dots\ r_{\prime}} \cdot \dfrac{}{n(n+1)\ n(n+1)^{3}\ n(n+1)^{m-2}\ n(n1)\ n(n+1)^{3}\ n(n+1)^{m-2}} \quad (3)'$$

CHAPTER III.

(I.) From among the infinite variety of developments which may be given to the general formulæ, a few will be here selected for the sake of illustration. Considering first the Ellipsoid, the formulæ appropriate to that surface will be found at page 26. In the first of these expressions, when $m = 0$ we have the true equation, if r_{-1} is assumed $= r^{-1}$,

$$\Sigma\left(\frac{x}{a}\right) = \frac{r}{\rho};$$

and the same value might be taken for m in the remaining formulæ, had not the suffix m, in the course of the investigation, been lowered by unity. We shall, however, in the following developments, in general assign to m the terms of the natural series, omitting zero.

The series will then be for the ellipsoid,

$$\frac{x}{a^3} + \frac{y}{b^3} + \frac{z}{c^3} = \frac{1}{\rho_1\, p}.$$

$$\frac{x}{a^5} + \frac{y}{b^5} + \frac{z}{c^5} = \frac{1}{\rho_2\, p_1\, p\, r_1}.$$

$$\frac{x}{a^7} + \frac{y}{b^7} + \frac{z}{c^7} = \frac{1}{\rho_3\, p_2\, p_1\, p\, r_2\, r_1}.$$

$$\frac{x}{a^9} + \frac{y}{b^9} + \frac{z}{c^9} = \frac{1}{\rho_4\, p_3\, p_2\, p_1\, p\, r_3\, r_2\, r_1}.$$

$$\frac{x}{a^{11}} + \frac{y}{b^{11}} + \frac{z}{c^{11}} = \frac{1}{\rho_5\, p_4\, p_3\, p_2\, p_1\, p\, r_4\, r_3\, r_2\, r_1}.$$

$$\cdots \qquad \cdots \qquad \qquad \cdots$$

$$\Sigma\left(\frac{x}{a^{2m+1}}\right) = \frac{1}{\rho_m\, p_{m-1}\, p_{m-2}\, \cdots\, p\, r_{m-1}\, r_{m-2}\, \cdots\, r_1}.$$

$$\frac{x^2}{a^2} + \frac{y^2}{b^2} + \frac{z^2}{c^2} = 1$$

$$\frac{x^2}{a^4} + \frac{y^2}{b^4} + \frac{z^2}{c^4} = \frac{1}{p^2}$$

$$\frac{x^2}{a^6} + \frac{y^2}{b^6} + \frac{z^2}{c^6} = \frac{1}{p^2 r_i^2}.$$

$$\frac{x^2}{a^8} + \frac{y^2}{a^8} + \frac{z^2}{a^8} = \frac{1}{p_i^2 p^2 r_i^2}.$$

$$\frac{x^2}{a^{10}} + \frac{y^2}{b^{10}} + \frac{z^2}{c^{10}} = \frac{1}{p_i^2 p^2 r_2^2 r_i^2}.$$

$$\frac{a^2}{a^{12}} + \frac{y^2}{a^{12}} + \frac{z^2}{c^{12}} = \frac{1}{p_2^2 p_i^2 p^2 r_2^2 r_i^2}.$$

$$\cdots \qquad \cdots \qquad \cdots$$

$$\Sigma\left(\frac{x^2}{a^{4m}}\right) = \frac{1}{p^2_{m-1} p^2_{m-2} \cdots p^2 r^2_{m-1} \cdots r_i^2}.$$

$$\Sigma\left(\frac{x^2}{a^{4m+2}}\right) = \frac{1}{p^2_{m-1} p^2_{m-2} \cdots p^2 r^2_m \cdots r_i^2}.$$

The expression (4), in page 26, will furnish the additional series, which may otherwise be derived by combining in pairs the terms of those which precede;

$$2\left\{\frac{xy}{a^3 b^3} + \frac{xz}{a^3 c^3} + \frac{yz}{b^3 c^3}\right\} = \frac{r_i^2 - \rho_i^2}{\rho_i^2 p^2 r_i^2}.$$

$$2\left\{\frac{xy}{a^5 b^5} + \frac{xz}{a^5 c^5} + \frac{yz}{b^5 c^5}\right\} = \frac{r_2^2 - \rho_2^2}{\rho_2^2 p_i^2 p^2 r_2^2 r_i^2}.$$

$$2\left\{\frac{xy}{a^7 b^7} + \frac{xz}{a^7 c^7} + \frac{yz}{b^7 c^7}\right\} = \frac{r_3^2 - \rho_3^2}{\rho_3^2 p_2^2 p_i^2 p^2 r_3^2 r_2^2 r_i^2}.$$

$$2\left\{\frac{xy}{a^9 b^9} + \frac{xz}{a^9 c^9} + \frac{yz}{b^9 c^9}\right\} = \frac{r_4^2 - \rho_4^2}{\rho_4^2 p_3^2 p_2^2 p_i^2 p^2 r_4^2 r_3^2 r_2^2 r_i^2}.$$

The formulæ of page 12 will give the absolute values of the consecutive radii and perpendiculars, in the following order,

$$r_1^2 = \frac{\frac{x^2}{a^4} + \cdots}{\frac{x^2}{a^6} + \cdots}; \quad \mu^2 = \frac{\frac{x^2}{a^2} + \cdots}{\frac{x^2}{a^4} + \cdots}; \quad \rho_1^2 = \frac{\frac{x^2}{a^4} + \cdots}{\left(\frac{x}{a^3} + \cdots\right)^2};$$

$$r_2^2 = \frac{\frac{x^2}{a^8} + \cdots}{\frac{x^2}{a^{10}} + \cdots}; \quad p_1^2 = \frac{\frac{x^2}{a^6} + \cdots}{\frac{x^2}{a^8} + \cdots}; \quad \rho_2^2 = \frac{\frac{x^2}{a^8} + \cdots}{\left(\frac{x}{a^5} + \cdots\right)^2};$$

$$r_3^2 = \frac{\frac{x^2}{a^{12}} + \cdots}{\frac{x^2}{a^{14}} + \cdots}; \quad p_2^2 = \frac{\frac{x^2}{a^{10}} + \cdots}{\frac{x^2}{a^{12}} + \cdots}; \quad \rho_3^2 = \frac{\frac{x^2}{a^{12}} + \cdots}{\left(\frac{x}{a^7} + \cdots\right)^2};$$

$$\cdots \qquad \cdots \qquad \cdots \qquad \cdots$$

the successive values of u will be found by combining the corresponding terms in the first and second columns; e.g.

$$u_2^4 = \frac{\Sigma\left(\frac{x^2}{a^{10}}\right)}{\Sigma\left(\frac{x^2}{a^{14}}\right)}.$$

(11.) Development for the surface,

$$\Sigma\left(\frac{x^3}{a^3}\right) = 1.$$

In this and the following developments for surfaces in which n is a positive integral number, in order to avoid repetition, we indicate each series by the number which marks its generating function in page 23.

(1.)—

$$m \;=\; 1, \quad \frac{x^3}{a^4} + \cdots \;=\; \frac{1}{\rho_i\, p}.$$

$$m \;=\; 2, \quad \frac{x^4}{a^{10}} + \cdots \;=\; \frac{1}{\rho_2\, p_i\, p^2\, r_i^2}.$$

$$m \;=\; 3, \quad \frac{x^8}{a^{22}} + \cdots \;=\; \frac{1}{\rho_3\, p_2\, p_i^2\, p^4\, r_2^2\, r_i^4}.$$

$$m \;=\; 4, \quad \frac{x^{16}}{a^{46}} + \cdots \;=\; \frac{1}{\rho_4\, p_3\, p_2^2\, p_i^4\, p^8\, r_3^2\, r_2^4\, r_i^8}.$$

$$m \;=\; 5, \quad \frac{x^{32}}{a^{94}} + \cdots \;=\; \frac{1}{\rho_5\, p_4\, p_3^2\, p_2^4\, p_i^8\, p^{16}\, r_4^2\, r_3^4\, r_2^9\, r_i^{16}}.$$

$$\cdots \qquad\qquad \cdots \qquad\qquad\qquad\qquad \cdots$$

(2.)—

$$m \;=\; 1, \quad \frac{x^4}{a^6} + \cdots \;=\; \frac{1}{p^2}.$$

$$m \;=\; 2, \quad \frac{x^8}{a^{13}} + \cdots \;=\; \frac{1}{p_i^3\, p^4\, r_i^4}.$$

$$m \;=\; 3, \quad \frac{x^{16}}{a^{43}} + \cdots \;=\; \frac{1}{p_2^3\, p_i^4\, p^8\, r_2^4\, r_i^8}.$$

$$m \;=\; 4, \quad \frac{x^{32}}{a^{90}} + \cdots \;=\; \frac{1}{p_3^3\, p_2^4\, p_i^8\, p^{16}\, r_3^4\, r_2^8\, r_i^{16}}.$$

$$m \;=\; 5, \quad \frac{x^{64}}{a^{186}} + \cdots \;=\; \frac{1}{p_4^2\, p_3^4\, p_2^8\, p_i^{16}\, p^{32}\, r_4^4\, r_3^8\, r_2^{16}\, r_i^{32}}.$$

(3.)—

$$m = 1, \quad \frac{x^6}{a^{12}} + \dots = \frac{1}{p^3 r_{/}^{\;3}}.$$

$$m = 2, \quad \frac{x^{12}}{a^{30}} + \dots = \frac{1}{p_{/}^{\;3}\, p^6\, r_2^{\;3}\, r_{/}^{\;6}}.$$

$$m = 3, \quad \frac{x^{24}}{a^{66}} + \dots = \frac{1}{p_2^{\;3}\, p_{/}^{\;6}\, p^{12}\, r_3^{\;3}\, r_2^{\;6}\, r_{/}^{\;12}}.$$

$$n = 4, \quad \frac{x^{48}}{a^{138}} + \dots = \frac{1}{p_3^{\;3}\, p_2^{\;6}\, p_{/}^{\;12}\, p^{24}\, r_4^{\;3}\, r_3^{\;6}\, r_2^{\;12}\, r_{/}^{\;24}}.$$

$$m = 5, \quad \frac{x^{96}}{a^{283}} + \dots = \frac{1}{p_4^{\;3}\, p_3^{\;6}\, p_2^{\;12}\, p_{/}^{\;24}\, p^{48}\, r_5^{\;3}\, r_4^{\;6}\, r_3^{\;12}\, r_2^{\;24}\, r_{/}^{\;48}}.$$

.. ...

(4.)—The formula (4), in page 23, will give for this surface the series,

$$\left\{\frac{x^2}{a^4} + \dots\right\}^3 - \left\{\left(\frac{x^2}{a^4}\right)^3 + \dots\right\} = \frac{r_{/}^{\;3} - \rho_{/}^{\;3}}{\rho_{/}^{\;3}\, p^3\, r_{/}^{\;3}}.$$

$$\left\{\frac{x^4}{a^{10}} + \dots\right\}^3 - \left\{\left(\frac{x^4}{a^{10}}\right)^3 + \dots\right\} = \frac{r_2^{\;3} - \rho_2^{\;3}}{\rho_2^{\;3}\, p_{/}^{\;3}\, p^6\, r_2^{\;3}\, r_{/}^{\;6}}.$$

$$\left\{\frac{x^8}{a^{22}} + \dots\right\}^3 - \left\{\left(\frac{x^8}{a^{22}}\right)^3 + \dots\right\} = \frac{r_3^{\;3} - \rho_3^{\;3}}{\rho_3^{\;3}\, p_2^{\;3}\, p_{/}^{\;6}\, p^{12}\, r_3^{\;3}\, r_2^{\;6}\, r_{/}^{\;12}}.$$

$$\left\{\frac{x^{16}}{a^{46}} + \dots\right\}^3 - \left\{\left(\frac{x^{16}}{a^{46}}\right)^3 + \dots\right\} = \frac{r_4^{\;3} - \rho_4^{\;3}}{\rho_4^{\;3}\, p_3^{\;3}\, p_2^{\;6}\, p_{/}^{\;12}\, p^{24}\, r_4^{\;3}\, r_3^{\;6}\, r_2^{\;12}\, r_{/}^{\;24}}.$$

$$\left\{\frac{x^{32}}{a^{94}} + \dots\right\}^3 - \left\{\left(\frac{x^{32}}{a^{94}}\right)^3 + \dots\right\} = \frac{r_5^{\;3} - \rho_5^{\;3}}{\rho_5^{\;3}\, p_4^{\;3}\, p_3^{\;6}\, p_2^{\;12}\, p_{/}^{\;24}\, p^{48}\, r_5^{\;3}\, r_4^{\;6}\, r_3^{\;12}\, r_2^{\;24}\, r_{/}^{\;48}}.$$

... ...

G

(III.) Development for the surfaee,

$$\Sigma \left(\frac{x^4}{a^4} \right) = 1.$$

(1.)—

$m = 1,\quad \dfrac{x^3}{a^5} + \ldots = \dfrac{1}{\rho_i\, p}.$

$m = 2,\quad \dfrac{a^9}{a^{17}} + \ldots = \dfrac{1}{\rho_2\, p_i\, p^3\, r_i{}^3}.$

$m = 3,\quad \dfrac{x^{27}}{a^{53}} + \ldots = \dfrac{1}{\rho_3\, p_2\, p_i{}^3\, p^9\, r_2{}^3\, r_i{}^9}.$

$m = 4,\quad \dfrac{x^{81}}{a^{161}} + \ldots = \dfrac{1}{\rho_4\, p_3\, p_2{}^3\, p_i{}^9\, p^{27}\, r_3{}^3\, r_2{}^9\, r_i{}^{27}}.$

$m = 5,\quad \dfrac{x^{243}}{a^{485}} + \ldots = \dfrac{1}{\rho_s\, p_4\, p_3{}^3\, p_2{}^9\, p_i{}^{27}\, p^{81}\, r_4{}^3\, r_3{}^9\, r_2{}^{27}\, r_i{}^{81}}.$

(2.)—

$m = 1,\quad \dfrac{x^6}{a^8} + \quad\;\; = \dfrac{1}{p^2}.$

$m = 2,\quad \dfrac{x^{18}}{a^{32}} + \ldots = \dfrac{1}{p_i{}^2\, p^6\, r_i{}^6}.$

$m = 3,\quad \dfrac{x^{54}}{a^{104}} + \ldots = \dfrac{1}{p_2{}^2\, p_i{}^6\, p^{18}\, r_2{}^6\, r_i{}^{18}}.$

$m = 4,\quad \dfrac{x^{162}}{a^{320}} + \ldots = \dfrac{1}{p_3{}^2\, p_2{}^6\, p_i{}^{18}\, p^{54}\, r_3{}^6\, r_2{}^{18}\, r_i{}^{54}}.$

$m = 5,\quad \dfrac{x^{480}}{a^{968}} + \ldots = \dfrac{1}{p_4{}^2\, p_3{}^6\, p_2{}^{18}\, p_i{}^{54}\, p^{162}\, r_4{}^6\, r_3{}^{18}\, r_2{}^{54}\, r_i{}^{162}}.$

(3.)—

$$m = 1, \quad \frac{a^{12}}{a^{20}} + \cdots = \frac{1}{p^4 \, r_i^{\,4}}.$$

$$m = 2, \quad \frac{x^{36}}{a^{68}} + \cdots = \frac{1}{p_i^{\,4} \, p^{12} \, r_2^{\,4} \, r_i^{\,12}}.$$

$$m = 3, \quad \frac{x^{108}}{a^{212}} + \cdots = \frac{1}{p_2^{\,4} \, p_i^{\,12} \, p^{36} \, r_3^{\,4} \, r_2^{\,12} \, r_i^{\,36}}.$$

$$m = 4, \quad \frac{x^{324}}{a^{644}} + \cdots = \frac{1}{p_3^{\,4} \, p_2^{\,12} \, p_i^{\,36} \, p^{108} \, r_4^{\,4} \, r_3^{\,12} \, r_2^{\,36} \, r_i^{\,108}}.$$

$$m = 5, \quad \frac{x^{972}}{a^{1940}} + \cdots = \frac{1}{p_4^{\,4} \, p_3^{\,12} \, p_2^{\,36} \, p_i^{\,108} \, p^{324} \, r_5^{\,4} \, r_4^{\,12} \, r_3^{\,36} \, r_2^{\,108} \, r_i^{\,324}}.$$

$$\cdots \qquad \qquad \cdots$$

(4.)—The formula (4) will give, in this case, the series,

$$\left\{ \frac{x^3}{a^5} + \cdots \right\}^4 - \left\{ \left(\frac{x^3}{a^5} \right)^4 + \cdots \right\} = \frac{r_i^{\,4} - \rho_i^{\,4}}{\rho_i^{\,4} \, p^4 \, r_i^{\,4}}.$$

$$\left\{ \frac{x^9}{a^{17}} + \cdots \right\}^4 - \left\{ \left(\frac{x^9}{a^{17}} \right)^4 + \cdots \right\} = \frac{r_2^{\,4} - \rho_2^{\,4}}{\rho_2^{\,4} \, p_i^{\,4} \, p^{12} \, r_2^{\,4} \, r_i^{\,12}}.$$

$$\left\{ \frac{x^{27}}{a^{53}} + \cdots \right\}^4 - \left\{ \left(\frac{x^{27}}{a^{53}} \right)^4 + \cdots \right\} = \frac{r_3^{\,4} - \rho_3^{\,4}}{\rho_3^{\,4} \, p_2^{\,4} \, p_i^{\,12} \, p^{36} \, r_3^{\,4} \, r_2^{\,12} \, r_i^{\,36}}.$$

$$\left\{ \frac{x^{81}}{a^{101}} + \cdots \right\}^4 - \left\{ \left(\frac{x^{81}}{a^{101}} \right)^4 + \cdots \right\} = \frac{r_4^{\,4} - \rho_4^{\,4}}{\rho_4^{\,4} \, p_3^{\,4} \, p_2^{\,12} \, p_i^{\,36} \, p^{108} \, r_4^{\,4} \, r_3^{\,12} \, r_2^{\,36} \, r_i^{\,108}}.$$

$$\left\{ \frac{x^{243}}{a^{485}} + \cdots \right\}^4 - \left\{ \left(\frac{x^{243}}{a^{485}} \right)^4 + \cdots \right\} = \frac{r_5^{\,4} - \rho_4^{\,4}}{\rho_5^{\,4} \, p_4^{\,4} \, p_3^{\,12} \, p_2^{\,36} \, p_i^{\,108} \, p^{324} \, r_5^{\,4} \, r_4^{\,12} \, r_3^{\,36} \, r_2^{\,108} \, r_i^{\,324}}.$$

$$\cdots \qquad \cdots \qquad \cdots$$

(IV.) Development for the surface,

$$\Sigma\left(\frac{x^5}{a^5}\right) = 1.$$

(1.)—

$m = 1, \quad \dfrac{x^4}{a^6} + \ldots = \dfrac{1}{\rho_{,}\,p}.$

$m = 2, \quad \dfrac{x^{16}}{a^{26}} + \ldots = \dfrac{1}{\rho_2\,p_{,}\,p^4\,r_{,}^4}.$

$m = 3, \quad \dfrac{x^{64}}{a^{106}} + \ldots = \dfrac{1}{\rho_3\,p_2\,p_{,}^4\,p^{16}\,r_2^4\,r_{,}^{16}}.$

$m = 4, \quad \dfrac{x^{256}}{a^{426}} + \ldots = \dfrac{1}{\rho_4\,p_3\,p_2^4\,p_{,}^{16}\,p^{64}\,r_3^4\,r_2^{16}\,r_{,}^{64}}.$

$m = 5, \quad \dfrac{x^{1024}}{a^{1706}} + \ldots = \dfrac{1}{\rho_5\,p_4\,p_3^4\,p_2^{16}\,p_{,}^{64}\,p^{256}\,r_4^4\,r_3^{16}\,r_2^{64}\,r_{,}^{256}}.$

$$\ldots$$

(2.)—

$m = 1, \quad \dfrac{x^8}{a^{10}} + \ldots = \dfrac{1}{p^2}.$

$m = 2, \quad \dfrac{x^{32}}{a^{50}} + \ldots = \dfrac{1}{p_{,}^2\,p^8\,r_{,}^8}.$

$m = 3, \quad \dfrac{x^{128}}{a^{210}} + \ldots = \dfrac{1}{p_2^2\,p_{,}^8\,p^{32}\,r_2^8\,r_{,}^{32}}.$

$m = 4, \quad \dfrac{x^{512}}{a^{850}} + \ldots = \dfrac{1}{p_3^2\,p_2^8\,p_{,}^{32}\,p^{128}\,r_3^8\,r_2^{32}\,r_{,}^{128}}.$

$m = 5, \quad \dfrac{x^{2048}}{a^{3410}} + \ldots = \dfrac{1}{p_4^2\,p_3^8\,p_2^{32}\,p_{,}^{128}\,p^{512}\,r_4^8\,r_3^{32}\,r_2^{128}\,r_{,}^{512}}.$

(3.)—

$$m = 1, \quad \frac{x^{20}}{a^{30}} + \ldots = \frac{1}{p^5 r_i^5}.$$

$$m = 2, \quad \frac{x^{80}}{a^{130}} + \ldots = \frac{1}{p_i^5 \, p^{20} \, r_2^5 \, r_i^{20}}.$$

$$m = 3, \quad \frac{x^{320}}{a^{530}} + \ldots = \frac{1}{p_2^5 \, p_i^{20} \, p^{80} \, r_3^5 \, r_2^{20} \, r_i^{80}}.$$

$$m = 4, \quad \frac{x^{1280}}{a^{2130}} + \ldots = \frac{1}{p_3^5 \, p_2^{20} \, p_i^{80} \, p^{320} \, r_4^5 \, r_3^{20} \, r_2^{80} \, r_i^{320}}.$$

$$m = 5, \quad \frac{x^{5120}}{a^{5530}} + \ldots = \frac{1}{p_4^5 \, p_3^{20} \, p_2^{80} \, p_i^{320} \, p^{1280} \, r_5^5 \, r_4^{20} \, r_3^{80} \, r_2^{320} \, r_i^{1280}}.$$

$$\ldots$$

(4.)—

$$\left\{\frac{x^4}{a^6} + \ldots\right\}^5 - \left\{\left(\frac{x^4}{a^6}\right)^5 + \ldots\right\} = \frac{r_i^5 - \rho_i^5}{\rho_i^5 \, p^5 \, r_i^5}.$$

$$\left\{\frac{x^{16}}{a^{26}} + \ldots\right\}^5 - \left\{\left(\frac{x^{16}}{a^{26}}\right)^5 + \ldots\right\} = \frac{r_2^5 - \rho_2^5}{\rho_2^5 \, p_i^5 \, p^{20} \, r_2^5 \, r_i^{20}}.$$

$$\left\{\frac{x^{64}}{a^{106}} + \ldots\right\}^5 - \left\{\left(\frac{x^{64}}{a^{106}}\right)^5 + \ldots\right\} = \frac{r_3^5 - \rho_3^5}{\rho_3^5 \, p_2^5 \, p_i^{20} \, p^{80} \, r_3^5 \, r_2^{20} \, r_i^{80}}.$$

$$\left\{\frac{x^{256}}{a^{426}} + \ldots\right\}^5 - \left\{\left(\frac{x^{256}}{a^{426}}\right)^5 + \ldots\right\} = \frac{r_4^5 - \rho_4^5}{\rho_4^5 \, p_3^5 \, p_2^{20} \, p_i^{80} \, p^{320} \, r_4^5 \, r_3^{20} \, r_2^{80} \, r^{320}}.$$

$$\left\{\frac{x^{1024}}{a^{1706}} + \ldots\right\}^5 - \left\{\left(\frac{x^{1024}}{a^{1706}}\right)^5 + \ldots\right\} = \frac{r_5^5 - \rho_5^5}{\rho_5^5 \, p_4^5 \, p_3^{20} \, p_2^{80} \, p_i^{320} \, p^{1280} \, r_5^5 \, r_4^{20} \, r_3^{80} \, r_2^{320} \, r_i}.$$

(V.)   Development of the formulæ for the surface,

$$\Sigma\left(\frac{x^{10}}{a^{10}}\right) \;=\; 1.$$

(1.)—

$$m = 1, \quad \frac{x^{9}}{a^{11}} + \dots \;=\; \frac{1}{\rho_{,}\,p}.$$

$$m = 2, \quad \frac{x^{81}}{a^{101}} + \dots \;=\; \frac{1}{\rho_{2}\,p_{,}\,p^{9}\,r_{,}^{9}}.$$

$$m = 3, \quad \frac{x^{729}}{a^{911}} + \dots \;=\; \frac{1}{\rho_{3}\,p_{2}\,p_{,}^{9}\,p^{81}\,r_{2}^{9}\,r_{,}^{81}}.$$

$$m = 4, \quad \frac{x^{6561}}{a^{8201}} + \dots \;=\; \frac{1}{\rho_{4}\,p_{3}\,p_{2}^{9}\,p_{,}^{81}\,p^{729}\,r_{3}^{9}\,r_{2}^{81}\,r_{,}^{729}}.$$

$$m = 5, \quad \frac{x^{59049}}{a^{73811}} \;=\; \frac{1}{\rho_{5}\,p_{4}\,p_{3}^{9}\,p_{2}^{81}\,p_{,}^{729}\,p^{6561}\,r_{4}^{9}\,r_{3}^{81}\,r_{2}^{729}\,r_{,}^{6561}}.$$

$$\dots \qquad\qquad\qquad \dots$$

(2.)—

$$m = 1, \quad \frac{x^{18}}{a^{20}} + \dots \;=\; \frac{1}{p^{2}}.$$

$$m = 2, \quad \frac{x^{162}}{a^{200}} + \dots \;=\; \frac{1}{p_{,}^{2}\,p^{18}\,r_{,}^{18}}.$$

$$m = 3, \quad \frac{x^{1458}}{a^{1820}} + \dots \;=\; \frac{1}{p_{2}^{2}\,p_{,}^{18}\,p^{162}\,r_{2}^{18}\,r_{,}^{162}}.$$

$$m = 4, \quad \frac{x^{13122}}{a^{16400}} + \dots \;=\; \frac{1}{p_{3}^{2}\,p_{2}^{18}\,p_{,}^{162}\,p^{1458}\,r_{3}^{18}\,r_{2}^{162}\,r_{,}^{1458}}.$$

$$m = 5, \quad \frac{x^{118098}}{a^{147020}} + \dots \;=\; \frac{1}{p_{4}^{2}\,p_{3}^{18}\,p_{2}^{162}\,p_{,}^{1458}\,p^{13122}\,r_{4}^{18}\,r_{3}^{162}\,r_{2}^{1458}\,r_{,}^{13122}}.$$

$$\dots \qquad \dots \qquad\qquad \dots$$

(3.)—

$$m = 1, \quad \frac{x^{90}}{a^{110}} + \ldots = \frac{1}{p^{10} \, r_{_i}^{10}}.$$

$$m = 2, \quad \frac{x^{810}}{a^{1010}} + \ldots = \frac{1}{p_{_i}^{10} \, p^{90} \, r_{_2}^{10} \, r_{_i}^{90}}.$$

$$m = 3, \quad \frac{x^{7290}}{a^{9110}} + \ldots = \frac{1}{p_{_2}^{10} \, p_{_i}^{90} \, p^{810} \, r_{_3}^{10} \, r_{_2}^{90} \, r_{_i}^{810}}.$$

$$m = 4, \quad \frac{x^{65610}}{a^{82010}} + \ldots = \frac{1}{p_{_3}^{10} \, p_{_2}^{90} \, p_{_i}^{810} \, p^{7290} \, r_{_4}^{10} \, r_{_3}^{90} \, r_{_2}^{810} \, r_{_i}^{7290}}.$$

$$m = 6, \quad \frac{x^{590490}}{a^{738110}} + \ldots = \frac{1}{p_{_4}^{10} \, p_{_3}^{90} \, p_{_2}^{810} \, p_{_i}^{7290} \, p^{65610} \, r_{_5}^{10} \, r_{_4}^{90} \, r_{_3}^{810} \, r_{_2}^{7290} \, r_{_i}^{65610}}.$$

$$\ldots$$

(4.)—

$$\left\{ \frac{x^9}{a^{11}} + \ldots \right\}^{10} - \left\{ \left(\frac{x^9}{a^{11}} \right)^{10} + \ldots \right\} = \frac{r_{_i}^{10} - \rho_{_i}^{10}}{\rho_{_i}^{10} \, p^{10} \, r_{_i}^{10}}.$$

$$\left\{ \frac{x^{81}}{a^{101}} + \ldots \right\}^{10} - \left\{ \left(\frac{x^{81}}{a^{101}} \right)^{10} + \ldots \right\} = \frac{r_{_2}^{10} - \rho_{_2}^{10}}{\rho_{_2}^{10} \, p_{_i}^{10} \, p^{90} \, r_{_2}^{10} \, r_{_i}^{90}}.$$

$$\left\{ \frac{x^{729}}{a^{911}} + \ldots \right\}^{10} - \left\{ \left(\frac{x^{729}}{a^{911}} \right)^{10} + \ldots \right\} = \frac{r_{_3}^{10} - \rho_{_3}^{10}}{\rho_{_3}^{10} \, p_{_2}^{10} \, p_{_i}^{90} \, p^{810} \, r_{_3}^{10} \, r_{_2}^{90} \, r_{_i}^{810}}.$$

$$\left\{ \frac{x^{6561}}{a^{8201}} + \ldots \right\}^{10} - \left\{ \left(\frac{x^{6561}}{a^{8101}} \right)^{10} + \ldots \right\} = \frac{r_{_4}^{10} - \rho_{_4}^{10}}{\rho_{_4}^{10} \, p_{_3}^{10} \, p_{_2}^{90} \, p_{_i}^{810} \, p^{7290} \, r_{_4}^{10} \, r_{_3}^{90} \, r_{_2}^{810} \, r_{_i}^{7290}}.$$

$$\left\{ \frac{x^{59049}}{a^{78811}} + \ldots \right\}^{10} - \left\{ \left(\frac{x^{59049}}{a^{78811}} \right)^{10} + \ldots \right\} = \frac{r_{_5}^{10} - \rho_{_5}^{10}}{\rho_{_5}^{10} \, p_{_4}^{10} \, p_{_3}^{90} \ldots p^{65610} \, r_{_5}^{10} \, r_{_4}^{90} \ldots r_{_i}^{65610}}.$$

(VI.) In each of the surfaces which have been considered, the absolute values of the consecutive radii and perpendiculars may be obtained from the forms in page 24. The developments of the general formulæ are susceptible of indefinite extension. The series, however, cannot here be conveniently extended ; nor is this necessary for the object of the treatise, since the examples already adduced will suffice to illustrate the power of the general forms. This division of the subject may, therefore, be con‑ cluded by annexing one or two illustrations derived from surfaces in which the indices are fractional or negative. For surfaces of which the characteristic index is integral and negative, the formulæ have been given at page 36; and general forms may be obtained expressing the corre‑ sponding relations when the indices are fractional. The expression of these relations does not, however, appear to be of much importance, since the adaptation of the original formulæ to any particular instance, or to any series of surfaces of any order, can present no difficulty, while it may, perhaps, be practically more convenient.

Before dismissing the surfaces with positive integral indices, it may be observed that the indices of $a\,b\,c$ are necessarily integral in every case in which m is integral and positive ; i.e., in all cases which here fall under our consideration : for,

$$\text{if } k = n-1, \quad \frac{(n-1)\overset{m}{-}1}{n-2} = \frac{k\overset{m}{-}1}{k-1} \;; \quad \text{which is always integral.}$$

$$\text{And,} \qquad \frac{n(n-1)\overset{m}{-}2}{n-2} = (n-1)\overset{m}{} + 2\frac{k\overset{m}{-}1}{k-1} \;;$$

which is consequently an integral number.

In the same way it may be seen, at page 36, that the indices are integers if n is *negative* and integral; and this will be self-evident, if we consider that in any case each of the indices in question represents the sum of a geometric series in which every term is separately an integer.

As a negative check upon the numerical accuracy of the several ex‑ pansions, it is useful to notice that in every case the sum of the indices on either side of the development must present an identity ; e.g. in (3), page 47, $738110 - 590490 = 147620$, which will be found to be the sum of the indices in the reciprocal equivalent. The general formulæ are, of course, subject to the same restriction of homogeneity.

(VII.) Development for the surface,

$$\Sigma\left(\frac{x^{\frac{1}{2}}}{a^{\frac{1}{2}}}\right) = 1.$$

The general formulæ, in this instance, are reduced to the following expressions:

$$\Sigma\left\{\frac{x^{\left(\frac{-1}{2}\right)^m}}{a^{\frac{m+2}{3} \cdot 2^{\frac{2-(-1)^m}{m}}}}\right\} = \frac{p_{m-2}^{\frac{1}{2}}\ p_{m-4}^{\frac{1}{4}}\ \ldots\ r_{m-1}^{\frac{1}{8}}\ r_{m-3}^{\frac{1}{8}}\ \ldots}{p_m\ p_{m-1}^{\frac{1}{4}}\ p_{m-3}^{\frac{1}{16}}\ p_{m-5}\ \ldots\ r_{m-2}^{\frac{1}{4}}\ r_{m-4}^{\frac{1}{16}}\ \ldots}. \tag{1}$$

$$\Sigma\left\{\frac{x^{\left(\frac{-1}{2}\right)^m}}{a^{\frac{m}{3} \cdot 2^{\frac{2-(-1)^m}{m}}}}\right\}^2 = \frac{p_{m-2}^{\frac{1}{4}}\ p_{m-4}^{\frac{1}{4}}\ \ldots\ r_{m-1}^{\frac{1}{4}}\ r_{m-3}^{\frac{1}{4}}\ \ldots}{p_{m-1}^{\frac{1}{2}}\ p_{m-3}^{\frac{1}{2}}\ p_{m-5}^{\frac{1}{2}}\ \ldots\ r_{m-2}^{\frac{1}{2}}\ r_{m-4}^{\frac{1}{2}}\ \ldots}. \tag{2}$$

$$\Sigma\left\{\frac{x^{\left(\frac{-1}{2}\right)^m}}{a^{\frac{m+2}{3} \cdot 2^{\frac{2-(-1)^m}{m}}}}\right\}^{\frac{1}{2}} = \frac{p_{m-2}^{\frac{1}{4}}\ p_{m-4}^{\frac{1}{16}}\ \ldots\ r_{m-1}^{\frac{1}{4}}\ r_{m-3}^{\frac{1}{16}}\ \ldots}{p_{m-1}^{\frac{1}{2}}\ p_{m-3}^{\frac{1}{8}}\ \ldots\ r_m^{\frac{1}{2}}\ r_{m-2}^{\frac{1}{8}}\ \ldots}. \tag{3}$$

(1.)—

$$m = 1, \quad \frac{1}{a_{\frac{1}{3}}^{\frac{1}{3}} x^{\frac{1}{2}}} + \ldots = \frac{1}{\rho_l\, p}.$$

$$m = 2, \quad \frac{x^{\frac{1}{2}}}{a^{\frac{1}{2}}} + \ldots = \frac{p^{\frac{1}{2}}\, r_l^{\frac{1}{4}}}{\rho_2\, p_l}.$$

$$m = 3, \quad \frac{1}{a^{\frac{5}{3}} x^{\frac{1}{2}}} + \ldots = \frac{\rho_l^{\frac{1}{4}}\, r_2^{\frac{1}{4}}}{\rho_3\, p_2\, p^{\frac{1}{4}}\, r_l^{\frac{1}{4}}}.$$

$$m = 4, \quad \frac{x^{\frac{1}{16}}}{a^{\frac{3}{16}}} + \ldots = \frac{p_2^{\frac{1}{2}}\, p^{\frac{1}{4}}\, r_3^{\frac{1}{2}}\, r_l^{\frac{1}{4}}}{\rho_4\, p_3\, p_l^{\frac{1}{4}}\, r_2^{\frac{1}{4}}}.$$

$$m = 5, \quad \frac{1}{a^{\frac{11}{3}} x^{\frac{1}{2}}} + \ldots = \frac{p_3^{\frac{1}{2}}\, p_l^{\frac{1}{8}}\, r_4^{\frac{1}{4}}\, r_2^{\frac{1}{2}}}{\rho_5\, p_4\, p_2^{\frac{1}{4}}\, p_{l_6}\, r_3^{\frac{1}{4}}\, r_{l_{16}}}.$$

... ...

(2.)—

$$m = 1, \quad \frac{1}{ax} + \ldots = \frac{1}{p^2}.$$

$$m = 2, \quad \frac{x^{\frac{1}{2}}}{a^{\frac{1}{2}}} + \ldots = \frac{p r_{\prime}}{p^2}.$$

$$m = 3, \quad \frac{1}{a^{\frac{1}{4}} x^{\frac{1}{2}}} + \ldots = \frac{p_{\prime}\, r_2}{p_2{}^2\, p_1^{\frac{1}{2}}\, r_{\prime}^{\frac{1}{2}}}.$$

$$m = 4, \quad \frac{x^{\frac{1}{2}}}{a^{\frac{2}{3}}} + \ldots = \frac{p_2\, p_1^{\frac{1}{2}}\, r_3\, r_{\prime}^{\frac{1}{4}}}{p_3{}^2\, p_1^{\frac{1}{3}}\, r_2^{\frac{1}{2}}}.$$

$$m = 5, \quad \frac{1}{a^{\frac{1}{16}} x^{\frac{1}{16}}} + \ldots = \frac{p_3\, p_1^{\frac{1}{4}}\, r_4\, r_2^{\frac{1}{2}}}{p_4{}^2\, p_2^{\frac{1}{2}}\, p_1^{\frac{1}{3}}\, r_3^{\frac{1}{2}}\, r_1^{\frac{1}{4}}}.$$

$$\ldots \qquad \ldots \qquad \ldots$$

(3.)—

$$m = 1, \quad \frac{1}{a^{\frac{1}{4}} x^{\frac{1}{4}}} + \ldots = \frac{1}{p^{\frac{1}{2}}\, r_{\prime}^{\frac{1}{8}}}.$$

$$m = 2, \quad \frac{x^{\frac{3}{8}}}{a^{\frac{3}{8}}} + \ldots = \frac{p^{\frac{1}{4}}\, r_{\prime}^{\frac{1}{4}}}{p_1^{\frac{1}{4}}\, r_2^{\frac{1}{2}}}.$$

$$m = 3, \quad \frac{1}{a^{\frac{1}{16}} x^{\frac{1}{16}}} + \ldots = \frac{p_{\prime}^{\frac{1}{4}}\, r_2^{\frac{1}{4}}}{p_2^{\frac{1}{2}}\, p_1^{\frac{1}{8}}\, r_3^{\frac{1}{2}}\, r_{\prime}^{\frac{1}{8}}}.$$

$$m = 4, \quad \frac{x^{\frac{7}{32}}}{a^{\frac{7}{32}}} + \ldots = \frac{p_2^{\frac{1}{4}}\, p_{10}^{\frac{1}{16}}\, r_3^{\frac{1}{4}}\, r_{\prime}^{\frac{1}{16}}}{p_3^{\frac{1}{2}}\, p_1^{\frac{1}{8}}\, r_4^{\frac{1}{2}}\, r_2^{\frac{1}{8}}}.$$

$$m = 5, \quad \frac{1}{a^{\frac{11}{64}} x^{\frac{11}{64}}} + \ldots = \frac{p_3^{\frac{1}{4}}\, p_1^{\frac{1}{16}}\, r_4^{\frac{1}{4}}\, r_2^{\frac{1}{16}}}{p_4^{\frac{1}{2}}\, p_2^{\frac{1}{4}}\, p_{32}\, r_5^{\frac{1}{2}}\, r_3^{\frac{1}{4}}\, r_{\prime 32}}.$$

(1.) —

$$\left\{\frac{1}{a^{\frac{2}{3}}\,x^{\frac{1}{2}}}+\ldots\right\}^{\frac{1}{2}}-\left\{\left(\frac{1}{a^{\frac{2}{3}}\,x^{\frac{1}{2}}}\right)^{\frac{1}{2}}+\ldots\right\}=\frac{r_{i}^{\frac{1}{2}}-\rho_{i}^{\frac{1}{2}}}{\rho_{i}^{\frac{1}{2}}\,p^{\frac{1}{2}}\,r_{i}^{\frac{1}{2}}}.$$

$$\left\{\frac{x^{\frac{1}{2}}}{a^{\frac{2}{3}}}+\ldots\right\}^{\frac{1}{2}}-\left\{\left(\frac{x^{\frac{1}{2}}}{a^{\frac{2}{3}}}\right)^{\frac{1}{2}}+\ldots\right\}=\frac{(r_{2}^{\frac{1}{2}}-\rho_{2}^{\frac{1}{2}})\,p^{\frac{1}{4}}\,r_{i}^{\frac{1}{4}}}{\rho_{2}^{\frac{1}{2}}\,p^{\frac{1}{2}}\,r_{2}^{\frac{1}{2}}}.$$

$$\left\{\frac{1}{a^{\frac{2}{3}}\,x^{\frac{1}{2}}}+\ldots\right\}^{\frac{1}{2}}-\left\{\left(\frac{1}{a^{\frac{2}{3}}\,x^{\frac{1}{2}}}\right)^{\frac{1}{2}}+\ldots\right\}=\frac{(r_{3}^{\frac{1}{2}}-\rho_{3}^{\frac{1}{2}})\,p_{i}^{\frac{1}{4}}\,r_{2}^{\frac{1}{4}}}{\rho_{3}^{\frac{1}{2}}\,p_{2}^{\frac{1}{2}}\,p^{\frac{1}{4}}\,r_{3}^{\frac{1}{2}}\,r_{i}^{\frac{1}{2}}}.$$

$$\left\{\frac{x^{\frac{1}{2}}}{a^{\frac{2}{3}}}+\ldots\right\}^{\frac{1}{2}}-\left\{\left(\frac{x^{\frac{1}{2}}}{a^{\frac{2}{3}}}\right)^{\frac{1}{2}}+\ldots\right\}=\frac{(r_{4}^{\frac{1}{2}}-\rho_{4}^{\frac{1}{2}})\,p_{2}^{\frac{1}{4}}\,p_{i}^{\frac{1}{4}}\,r_{3}^{\frac{1}{4}}\,r_{i}^{\frac{1}{8}}}{\rho_{4}^{\frac{1}{2}}\,p_{3}^{\frac{1}{2}}\,p_{i}^{\frac{1}{2}}\,r_{4}^{\frac{1}{2}}\,r_{2}^{\frac{1}{4}}}.$$

$$\left\{\frac{1}{a^{\frac{2}{3}}\,x^{\frac{1}{2}}}+\ldots\right\}^{\frac{1}{2}}-\left\{\left(\frac{1}{a^{\frac{2}{3}}\,x^{\frac{1}{2}}}\right)^{\frac{1}{2}}+\ldots\right\}=\frac{(r_{5}^{\frac{1}{2}}-\rho_{5}^{\frac{1}{2}})\,p_{3}^{\frac{1}{4}}\,p_{i}^{\frac{1}{16}}\,r_{4}^{\frac{1}{4}}\,r_{2}^{\frac{1}{16}}}{\rho_{5}^{\frac{1}{2}}\,p_{4}^{\frac{1}{2}}\,p_{2}^{\frac{1}{2}}\,p_{i}^{\frac{1}{2}}\,r_{5}^{\frac{1}{4}}\,r_{3}^{\frac{1}{2}}\,r_{i}^{\frac{1}{2}}}.$$

(VIII.) This surface presents the following geometrical peculiarity, derived from the development of (2).

When $m = 2$, it has been shewn that,

$$\Sigma\left(\frac{a^{\frac{1}{2}}}{n^{\frac{1}{2}}}\right)\;=\;\frac{pr_{i}}{p_{i}^{2}};$$

consequently, $\quad\dfrac{p r_{i}}{p_{i}^{2}}\;=\;1;\quad$ or, $p_{i}^{2}\;=\;pr_{i}:$

so that, in this instance, p_{i} is a mean proportional to p and r_{i}, and is therefore equal to the circle-ordinate u, (page 3).

In figure (2), let

$$\mathrm{OP}\;=\;r\,;\quad\mathrm{OP}_{i}\;=\;r_{i}\,;\quad\mathrm{OP}_{2}\;=\;r_{2}\,;$$

$$\mathrm{OQ}\;=\;p\,;\quad\mathrm{OQ}_{i}\;=\;p_{i}\,;\quad\mathrm{OQ}_{2}\;=\;p_{2}\,;$$

$$\therefore\;\mathrm{OQ}_{i}^{2}\;=\;\mathrm{OQ}\,.\,\mathrm{OP}_{i}\,;$$

and, since the law of derivation is invariable for all the consecutive points $\mathrm{Q}\,\mathrm{Q}_{i}\,\mathrm{Q}_{2}\,\ldots\,\mathrm{P}\,\mathrm{P}_{i}\,\mathrm{P}_{2}\,\ldots$ it follows that, $\mathrm{OQ}_{2}^{2}\;=\;\mathrm{OQ}_{i}\,.\,\mathrm{OP}_{2}$, and, generally, $\mathrm{OQ}_{m}^{2}\;=\;\mathrm{OQ}_{m-1}\,.\,\mathrm{OP}_{m}:$

$$\therefore\;p_{m}^{2}\;=\;p_{m-1}\,r_{m}.$$

(IX.) By employing the relation established in the preceding section, the developments admit of simplification, and may be expressed in the forms which follow.

(1.)—

$$m = 1, \quad \frac{1}{a^{\frac{3}{2}} x^{\frac{1}{2}}} + \ldots = \frac{1}{\rho_1\, \rho}.$$

$$m = 2, \quad \frac{x^{\frac{1}{2}}}{a^{\frac{3}{2}}} + \ldots = \frac{1}{\rho_2}.$$

$$m = 3, \quad \frac{1}{a^{\frac{5}{2}} x^{\frac{1}{2}}} + \ldots = \frac{1}{\rho_3\, p_1^{\frac{1}{2}}}.$$

$$m = 4, \quad \frac{x^{\frac{1}{2}}}{a^{\frac{5}{2}}} + \ldots = \frac{1}{\rho_4\, r_2^{\frac{1}{2}}}.$$

$$m = 5, \quad \frac{1}{a^{\frac{7}{2}} x^{\frac{1}{2}}} + \ldots = \frac{1}{\rho_5\, r_3^{\frac{1}{2}} p_1^{\frac{1}{2}}}.$$

$$\ldots \qquad \ldots \qquad \ldots$$

(2.)—

$$m = 1, \quad \frac{1}{ax} + \ldots = \frac{1}{p^2}.$$

$$m = 2, \quad \frac{x^{\frac{1}{2}}}{a^{\frac{1}{2}}} + \ldots = 1.$$

$$m = 3, \quad \frac{1}{a^{\frac{1}{2}} x^{\frac{1}{2}}} + \ldots = \frac{1}{p_1}.$$

$$m = 4, \quad \frac{x^{\frac{1}{2}}}{a^{\frac{1}{2}}} + \ldots = \frac{1}{r_2^{\frac{1}{2}}}.$$

$$m = 5, \quad \frac{1}{a^{\frac{1}{2}} x^{\frac{1}{2}}} + \ldots = \frac{1}{r_3^{\frac{1}{2}} p_1^{\frac{1}{2}}}.$$

(3.)—

$$m = 1, \quad \frac{1}{a^{\frac{1}{2}} x^{\frac{1}{2}}} + \cdots = \frac{1}{p_{\prime}}.$$

$$m = 2, \quad \frac{x^{\frac{1}{2}}}{a^{\frac{1}{2}}} + \cdots = \frac{1}{r_2^{\frac{1}{2}}}.$$

$$m = 3, \quad \frac{1}{a^{\frac{1}{8}} x^{\frac{1}{8}}} + \cdots = \frac{1}{r_3^{\frac{1}{2}} p_{\prime}^{\frac{1}{2}}}.$$

$$m = 4, \quad \frac{x^{\frac{1}{4}}}{a^{\frac{3}{4}}} + \cdots = \frac{1}{r_4^{\frac{1}{2}} r_2^{\frac{1}{2}}}.$$

$$m = 5, \quad \frac{1}{a^{\frac{11}{16}} x^{\frac{5}{16}}} + \cdots = \frac{1}{r_5^{\frac{1}{2}} r_3^{\frac{1}{2}} p_{\prime}^{\frac{1}{16}}}.$$

$$\cdots \qquad \cdots \qquad \cdots$$

(4.)—

$$\left\{ \frac{1}{a^{\frac{1}{2}} x^{\frac{1}{2}}} + \cdots \right\}^{\frac{1}{2}} - \left\{ \left(\frac{1}{a^{\frac{1}{2}} x^{\frac{1}{2}}} \right)^{\frac{1}{2}} + \cdots \right\} = \frac{r_{\prime}^{\frac{1}{2}} - \rho_{\prime}^{\frac{1}{2}}}{\rho_{\prime}^{\frac{1}{2}}} \cdot \frac{1}{p_{\prime}}.$$

$$\left\{ \frac{x^{\frac{1}{2}}}{a^{\frac{1}{2}}} + \cdots \right\}^{\frac{1}{2}} - \left\{ \left(\frac{x^{\frac{1}{2}}}{a^{\frac{1}{2}}} \right)^{\frac{1}{2}} + \cdots \right\} = \frac{r_{2}^{\frac{1}{2}} - \rho_{2}^{\frac{1}{2}}}{\rho_{2}^{\frac{1}{2}}} \cdot \frac{\rho_{\prime}^{\frac{1}{2}}}{p_2}.$$

$$\left\{ \frac{1}{a^{\frac{1}{8}} x^{\frac{1}{8}}} + \cdots \right\}^{\frac{1}{2}} - \left\{ \left(\frac{1}{a^{\frac{1}{8}} x^{\frac{1}{8}}} \right)^{\frac{1}{2}} + \cdots \right\} = \frac{r_{3}^{\frac{1}{2}} - \rho_{3}^{\frac{1}{2}}}{\rho_{3}^{\frac{1}{2}}} \cdot \frac{\rho_{2}^{\frac{1}{2}}}{p_3 \, p_{\prime}^{\frac{1}{2}}}.$$

$$\left\{ \frac{x^{\frac{1}{16}}}{a^{\frac{3}{16}}} + \cdots \right\}^{\frac{1}{2}} - \left\{ \left(\frac{x^{\frac{1}{16}}}{a^{\frac{3}{16}}} \right)^{\frac{1}{2}} + \cdots \right\} = \frac{r_{4}^{\frac{1}{2}} - \rho_{4}^{\frac{1}{2}}}{\rho_{4}^{\frac{1}{2}}} \cdot \frac{\rho_{3}^{\frac{1}{2}} \, p_{\prime}^{\frac{1}{6}}}{p_4 \, p_2^{\frac{1}{2}}}.$$

$$\left\{ \frac{1}{a^{\frac{11}{32}} x^{\frac{5}{32}}} + \cdots \right\}^{\frac{1}{2}} - \left\{ \left(\frac{1}{a^{\frac{11}{32}} x^{\frac{5}{32}}} \right)^{\frac{1}{2}} + \cdots \right\} = \frac{r_{5}^{\frac{1}{2}} - \rho_{5}^{\frac{1}{2}}}{\rho_{5}^{\frac{1}{2}}} \cdot \frac{p_{4}^{\frac{1}{2}} \, p_{2}^{\frac{1}{6}}}{p_5 \, p_2^{\frac{1}{2}} \, p_{\prime}^{\frac{1}{16}}}.$$

An examination of the equations will shew that, in the instance of this surface, owing to the singular relation $p_m{}^2 = p_{m-1}\, r_m$, the development of (3) is included in that of (2); and that this connexion must exist will be seen by reference to the general formulæ as adapted to the surface under consideration, page 49, in which, if $m + 2$ is substituted in (2), while m is written in the formula (3), both expressions will furnish coincident results.

(X.) In illustration of the formulæ for negative indices, as given in page 36, we may first select the surface,

$$\Sigma \left(\frac{a}{x} \right) \;=\; 1.$$

(1.)—

$$m = 1, \quad \frac{1}{x^2} + \ldots = \frac{1}{p_i\, p}.$$

$$m = 2, \quad \frac{x^4}{a^2} + \ldots = \frac{p^2\, r_i{}^2}{p_2\, p_i}.$$

$$m = 3, \quad \frac{a^2}{x^8} + \ldots = \frac{p_i{}^2\, r_2{}^2}{p_3\, p_2\, p^4\, r_i{}^4}.$$

$$m = 4, \quad \frac{x^{16}}{a^6} + \ldots = \frac{p_2{}^2\, p^8\, r_3{}^2\, r_i{}^8}{p_4\, p_3\, p_i{}^4\, r_2{}^4}.$$

$$m = 5, \quad \frac{a^{10}}{x^{32}} + \ldots = \frac{p_3{}^2\, p_i{}^8\, r_4{}^2\, r_2{}^8}{p_5\, p_4\, p_2{}^4\, p^{16}\, r_3{}^4\, r_i{}^{16}}.$$

$$m = 6, \quad \frac{x^{64}}{a^{22}} + \ldots = \frac{p_4{}^2\, p_2{}^8\, p^{32}\, r_5{}^2\, r_3{}^8\, r_i{}^{32}}{p_6\, p_5\, p_3{}^4\, p_i{}^{10}\, r_4{}^4\, r_2{}^{16}}.$$

$$m = 7, \quad \frac{a^{42}}{x^{128}} + \ldots = \frac{p_5{}^2\, p_3{}^8\, p_i{}^{32}\, r_6{}^2\, r_4{}^8\, r_2{}^{32}}{p_7\, p_6\, p_4{}^4\, p_2{}^{16}\, p^{64}\, r_5{}^4\, r_3{}^{16}\, r_i{}^{64}}.$$

$$m = 8, \quad \frac{x^{256}}{a^{86}} + \ldots = \frac{p_6{}^2\, p_4{}^8\, p_2{}^{32}\, p^{128}\, r_7{}^2\, r_5{}^8\, r_3{}^{32}\, r_i{}^{128}}{p_8\, p_7\, p_5{}^4\, p_3{}^{16}\, p_i{}^{64}\, r_6{}^4\, r_4{}^{16}\, r_2{}^{64}}.$$

In these developments the indices shew some curious relations. For example, taking any term in which m is *even*, the difference of its indices gives the index of the *numerator* in the following term ; but, if we consider any term in which m is *uneven*, the difference of the indices will be the index of the *denominator* in the expression which immediately succeeds. On the second side of each equation in which m is even, the sum of the indices in the numerator is always double of the sum of the indices in the denominator. The general formulæ will establish these relations without difficulty.

(2.)—

$$m = 1, \quad \frac{a^2}{x^4} + \dots = \frac{1}{p^2}.$$

$$m = 2, \quad \frac{x^8}{a^2} + \dots = \frac{p^4 r_i^4}{p_i^3}.$$

$$m = 3, \quad \frac{a^6}{x^{16}} + \dots = \frac{p_i^4 r_2^4}{p_2^2 p^8 r_i^5}.$$

$$m = 4, \quad \frac{x^{32}}{a^{10}} + \dots = \frac{p_2^4 p^{16} r_3^4 r_i^{16}}{p_3^2 p_i^8 r_2^8}.$$

$$m = 5, \quad \frac{a^{02}}{x^{64}} + \dots = \frac{p_3^4 p_i^{16} r_4^4 r_2^{16}}{p_4^2 p_2^8 p^{32} r_3^8 r^{32}}.$$

$$m = 6, \quad \frac{x^{128}}{a^{42}} + \dots = \frac{p_4^4 p_2^{16} p_i^{64} r_5^4 r_3^{16} r_i^{64}}{p_5^2 p_3^8 p^{32} r_4^8 r_2^{32}}.$$

$$m = 7, \quad \frac{a^{86}}{x^{256}} + \dots = \frac{p_5^4 p_3^{16} p_i^{64} r_6^4 r_4^{16} r_2^{64}}{p_6^2 p_4^8 p_2^{32} p^{128} r_5^8 r_3^{32} r_i^{128}}.$$

$$m = 8, \quad \frac{x^{512}}{a^{170}} + \dots = \frac{p_6^4 p_4^{16} p_2^{64} p^{256} r_7^4 r_5^{16} r_3^{64} r_i^{256}}{p_7^2 p_5^8 p_3^{32} p_i^{128} r_6^8 r_4^{32} r_2^{128}}.$$

(3.)—

$$m = 1, \quad x^2 \;+\; \ldots \;=\; pr_{/}.$$

$$m = 2, \quad \frac{a^2}{x^4} \;+\; \ldots \;=\; \frac{p_{/}\, r_2}{p^2\, r_{/}^2}.$$

$$m = 3, \quad \frac{x^8}{a^2} \;+\; \;=\; \frac{p_2\, p^4\, r_3\, r_{/}^4}{p_{/}^2\, r_2^2}.$$

$$m = 4, \quad \frac{a^6}{x^{16}} \;+\; \ldots \;=\; \frac{p_3\, p_{/}^4\, r_4\, r_2^4}{p_2^2\, p^8\, r_3^2\, r_{/}^8}.$$

$$m = 5, \quad \frac{x^{32}}{a^{10}} \;+\; \ldots \;=\; \frac{p_4\, p_2^4\, p^{16}\, r_5\, r_3^4\, r_{/}^{16}}{p_3^2\, p_{/}^8\, r_4^2\, r_2^8}.$$

$$m = 6, \quad \frac{a^{22}}{x^{64}} \;+\; \;=\; \frac{p_5\, p_3^4\, p_{/}^{16}\, r_6\, r_4^4\, r_2^{16}}{p_4^2\, p_2^8\, p^{32}\, r_5^2\, r_3^8\, r_{/}^{32}}.$$

$$m = 7, \quad \frac{x^{128}}{a^{42}} \;+\; \ldots \;=\; \frac{p_6\, p_4^4\, p_2^{16}\, p^{64}\, r_7\, r_5^4\, r_3^{16}\, r_{/}^{64}}{p_5^2\, p_3^8\, p_{/}^{32}\, r_6^2\, r_4^8\, r_2^{32}}.$$

$$m = 8, \quad \frac{a^{86}}{x^{256}} \;+\; \ldots \;=\; \frac{p_7\, p_5^4\, p_3^{16}\, p_{/}^{64}\, r_8\, r_6^4\, r_4^{16}\, r_2^{64}}{p_6^2\, p_4^8\, p_2^{32}\, p^{128}\, r_7^2\, r_5^8\, r_3^{32}\, r_{/}^{128}}.$$

$$\ldots \qquad \ldots \qquad\qquad \ldots$$

(XI.) It appears from the last development that the circle-ordinate v, in this surface, is equal to r, the radius vector at the first point of contact. In consequence of this singularity we shall have, by attending to the principle of uniformity which governs the position of the consecutive points on the surface, fig. 2, $v_m = r_m$;

$$\therefore \quad r_m^2 \;=\; p_m\, r_{m+1} :$$

in which it is to be understood that r and p are indicated by r_0 and p_0. This peculiar relation will have the effect of reducing the preceding expressions to others which involve only the lines $r\; r_{/}\; r_2\; r_3\; \ldots\; r_m$; in the form which follows.

(XII.) Reduction of the developments for the surface,

$$\Sigma \left(\frac{a}{x}\right) \;=\; 1.$$

(1.)—

$$m = 1, \quad \frac{1}{x^2} + \quad = \quad \frac{r_1}{\rho_1}\,\frac{1}{r^2}.$$

$$m = 2, \quad \frac{x^4}{a^2} + \ldots \quad = \quad \frac{r_2}{\rho_2}\,\frac{r^4}{r_1^2}.$$

$$m = 3, \quad \frac{a^2}{x^5} + \quad = \quad \frac{r_3}{\rho_3}\,\frac{r_1^4}{r_2^2\, r^8}.$$

$$m = 4, \quad \frac{x^{16}}{a^6} + \ldots \quad = \quad \frac{r_3}{\rho_4}\,\frac{r_2^4\, r^{16}}{r_3^2\, r_1^8}.$$

$$m = 5, \quad \frac{a^{10}}{x^{32}} + \quad = \quad \frac{r_5}{\rho_5}\,\frac{r_3^4\, r_1^{16}}{r_4^2\, r_2^8\, r^{32}}.$$

...   ...

(2.)—

$$m = 1, \quad \frac{a^2}{x^4} + \ldots \quad = \quad \frac{r_1^2}{r^4}.$$

$$m = 2, \quad \frac{x^8}{a^2} + \ldots \quad = \quad \frac{r_2^2\, r^8}{r_1^4}.$$

$$m = 3, \quad \frac{a^6}{x^{16}} + \ldots \quad = \quad \frac{r_3^2\, r_1^8}{r_2^4\, r^{16}}.$$

$$m = 4, \quad \frac{x^{32}}{a^{10}} + \ldots \quad = \quad \frac{r_4^2\, r_2^8\, r^{32}}{r_3^4\, r_1^{16}}.$$

$$m = 5, \quad \frac{a^{22}}{x^{64}} + \ldots \quad = \quad \frac{r_5^2\, r_3^8\, r_1^{32}}{r_4^4\, r_2^{16}\, r^{64}}.$$

(3.)—

$$m = 1, \quad x^2 \quad + \ldots \quad = \quad r^2.$$

$$m = 2, \quad \frac{a^2}{x^4} \quad + \quad \quad = \quad \frac{r_i^2}{r^4}.$$

$$m = 3, \quad \frac{x^8}{a^2} \quad + \ldots \quad = \quad \frac{r_2^2 \, r^8}{r_i^4}.$$

$$m = 4, \quad \frac{a^6}{x^{16}} \quad + \ldots \quad = \quad \frac{r_3^2 \, r_i^8}{r_2^4 \, r^{16}}.$$

$$m = 5, \quad \frac{x^{32}}{a^{10}} \quad + \ldots \quad = \quad \frac{r_4^2 \, r_2^8 \, r_i^{32}}{r_3^4 \, r_i^{16}}.$$

$$\ldots \qquad \ldots \qquad\qquad \ldots$$

(XIII.)　Development for the surface,

$$\Sigma \left(\frac{a^2}{x^3} \right) = 1.$$

(1.)—

$$m = 1, \quad \frac{a}{x^3} \quad + \ldots \quad = \quad \frac{1}{\rho_i \, p}.$$

$$m = 2, \quad \frac{x^9}{a^6} \quad + \ldots \quad = \quad \frac{p^3 \, r_i^3}{\rho_2 \, p_i}.$$

$$m = 3, \quad \frac{a^{13}}{x^{27}} \quad + \ldots \quad = \quad \frac{p_i^3 \, r_2^3}{\rho_3 \, p_2 \, p^9 \, r_i^9}.$$

$$m = 4, \quad \frac{x^{81}}{a^{41}} \quad + \ldots \quad = \quad \frac{p_2^3 \, p^{27} \, r_3^3 \, r_i^{27}}{\rho_4 \, p_3 \, p_i^9 \, r_2^9}.$$

$$\ldots \qquad\qquad \ldots$$

$$m = 8, \quad \frac{x^{6561}}{a^{3281}} \quad + \ldots \quad = \quad \frac{p_6^3 \, p_4^{27} \, p_2^{243} \, p^{2187} \, r_7^3 \, r_5^{27} \, r_3^{243} \, r_i^{2187}}{\rho_8 \, p_7 \, p_5^9 \, p_3^{81} \, p_i^{729} \, r_6^9 \, r_4^{81} \, r_2^{729}}.$$

(2.)—

$$m = 1, \quad \frac{a^4}{x^8} + \quad = \frac{1}{p^2}.$$

$$m = 2, \quad \frac{x^{18}}{a^8} + \dots = \frac{p^6 r^6}{p_i^3}.$$

$$m = 3, \quad \frac{a^{28}}{x^{54}} + \dots = \frac{p_i^6 r_2^6}{p_2^2 p^{18} r_i^{18}}.$$

$$m = 4, \quad \frac{x^{162}}{a^{80}} + \dots = \frac{p_2^6 p^{54} r_3^6 r_i^{54}}{p_3^2 p_i^{18} r_2^{18}}.$$

$$\dots \qquad \dots \qquad\qquad \dots$$

$$m = 8, \quad \frac{x^{13122}}{a^{6560}} + \dots = \frac{p_6^6 p_4^{54} p_2^{486} p^{4374} r_7^6 r_5^{54} r_3^{486} r^{4374}}{p_7^2 p_5^{18} p_3^{162} p_i^{1458} r_6^{18} r_4^{162} r_2^{1458}}.$$

$$\dots \qquad \dots \qquad\qquad \dots$$

(3.)—

$$m = 1, \quad \frac{x^6}{a^2} + \dots = p^2 r_i^3.$$

$$m = 2, \quad \frac{a^{10}}{x^{18}} + \dots = \frac{p_i^3 r_2^2}{p_6 r_{i6}}.$$

$$m = 3, \quad \frac{x^{54}}{a^{26}} + \dots = \frac{p_2^2 p^{18} r_3^2 r_i^{18}}{p_i^6 r_2^6}.$$

$$m = 4, \quad \frac{a^{82}}{x^{162}} + \dots = \frac{p_3^2 p_i^{18} r_4^2 r_2^{18}}{p_2^6 p^{54} r_3^6 r_i^{54}}.$$

$$\dots \qquad \dots \qquad\qquad \dots$$

$$m = 8, \quad \frac{a^{6562}}{x^{13122}} + \dots = \frac{p_7^2 p_5^{18} p_3^{162} p_i^{1458} r_8^2 r_6^{18} r_4^{162} r_2^{1458}}{p_6^6 p_4^{54} p_2^{486} p^{4374} r_7^6 r_5^{54} r_3^{486} r_i^{4374}}.$$

(XIV.)   Development for the surface,

$$\Sigma\left(\frac{a^3}{x^3}\right) \;=\; 1.$$

(1.)—

$m = 1,\quad \dfrac{a^2}{x^4} + \ldots = \dfrac{1}{\rho_{_1}\,p}.$

$m = 2,\quad \dfrac{x^{16}}{a^{10}} + \ldots = \dfrac{p^4\, r_{_1}^4}{\rho_2\, p_{_1}}.$

$m = 3,\quad \dfrac{a^{38}}{x^{64}} + \ldots = \dfrac{p_{_1}^4\, r_{_2}^4}{\rho_3\, p_2\, p^{16}\, r_{_1}^{16}}.$

$m = 4,\quad \dfrac{x^{256}}{a^{154}} + \ldots = \dfrac{p_{_2}^4\, p^{64}\, r_{_3}^4\, r_{_1}^{64}}{\rho_4\, p_3\, p_{_1}^{16}\, r_{_2}^{16}}.$

$m = 5,\quad \dfrac{a^{614}}{x^{1024}} + \ldots = \dfrac{p_{_3}^4\, p_{_1}^{64}\, r_{_4}^4\, r_{_2}^{64}}{\rho_5\, p_4\, p_2^{16}\, p^{256}\, r_{_3}^{16}\, r_{_1}^{256}}.$

...   .

(2.)—

$m = 1,\quad \dfrac{a^6}{x^8} + \ldots = \dfrac{1}{p^2}.$

$m = 2,\quad \dfrac{x^{32}}{a^{18}} + \ldots = \dfrac{p^8\, r^8}{p_{_1}^2}.$

$m = 3,\quad \dfrac{a^{78}}{x^{128}} + \ldots = \dfrac{p_{_1}^8\, r_{_2}^8}{p_{_2}^2\, p^{32}\, r^{32}}.$

$m = 4,\quad \dfrac{x^{512}}{a^{306}} + \ldots = \dfrac{p_{_2}^8\, p^{128}\, r_{_3}^8\, r_{_1}^{128}}{p_{_3}^2\, p_{_1}^{32}\, r_{_2}^{32}}.$

$m = 5,\quad \dfrac{a^{1230}}{x^{2048}} + \ldots = \dfrac{p_{_3}^8\, p_{_1}^{128}\, r_{_4}^8\, r_{_2}^{128}}{p_{_4}^2\, p_{_2}^{32}\, p^{513}\, r_{_3}^{32}\, r_{_1}^{512}}.$

(3.)—

$$m = 1, \quad \frac{x^{12}}{a^6} + \cdots = p^3\, r_i^3.$$

$$m = 2, \quad \frac{a^{30}}{x^{48}} + \cdots = \frac{p_i^{\,3}\, r_2^{\,3}}{p^{12}\, r_i^{\,12}}.$$

$$m = 3, \quad \frac{x^{192}}{a^{114}} + \cdots = \frac{p_2^{\,3}\, p_i^{\,48}\, r_3^{\,3}\, r_i^{\,48}}{p_i^{\,12}\, r_2^{\,12}}.$$

$$m = 4, \quad \frac{a^{462}}{x^{768}} + \cdots = \frac{p_3^{\,3}\, p_i^{\,48}\, r_4^{\,3}\, r_2^{\,48}}{p_2^{\,12}\, p_i^{\,192}\, r_3^{\,12}\, r_i^{\,192}}.$$

$$m = 5, \quad \frac{x^{3072}}{a^{1843}} + \cdots = \frac{p_4^{\,3}\, p_2^{\,48}\, p^{768}\, r_5^{\,3}\, r_3^{\,48}\, r_i^{\,768}}{p_3^{\,12}\, p_i^{\,192}\, r_4^{\,12}\, r_2^{\,192}}.$$

...   ...

(XV.) Development for the surface,

$$\Sigma\left(\frac{a^{\frac{1}{2}}}{x^{\frac{1}{2}}}\right) = 1.$$

(1.)—

$$m = 1, \quad \frac{1}{a^{\frac{1}{2}}\, x^{\frac{1}{2}}} + \cdots = \frac{1}{\rho_i\, p}.$$

$$m = 2, \quad \frac{x^{\frac{1}{2}}}{a^{\frac{1}{2}}} + \cdots = \frac{p^{\frac{1}{2}}\, r^{\frac{1}{2}}}{\rho_2\, p_i}.$$

$$m = 3, \quad \frac{1}{a^{\frac{1}{2}}\, x^{\frac{7}{2}}} + \cdots = \frac{p_i^{\,\frac{1}{2}}\, r_2^{\,\frac{1}{2}}}{\rho_3\, \rho_2\, p_i^{\,\frac{1}{2}}\, r_i^{\,\frac{1}{2}}}.$$

$$m = 4, \quad \frac{x^{\frac{11}{2}}}{a^{\frac{23}{2}}} + \cdots = \frac{p_2^{\,\frac{1}{2}}\, p^{\frac{7}{2}}\, r_3^{\,\frac{1}{2}}\, r^{\frac{7}{2}}}{\rho_4\, \rho_3\, p_i^{\,\frac{1}{2}}\, r_2^{\,\frac{1}{2}}}.$$

$$m = 5, \quad \frac{a^{\frac{31}{2}}}{x^{\frac{37}{2}}} + \cdots = \frac{p_3^{\,\frac{1}{2}}\, p_i^{\,\frac{7}{2}}\, r_4^{\,\frac{1}{2}}\, r_2^{\,\frac{7}{2}}}{\rho_5\, \rho_4\, p_2^{\,\frac{1}{2}}\, p^{\frac{11}{2}}\, r_3^{\,\frac{1}{2}}\, r_i^{\,\frac{11}{2}}}.$$

(2.)—

$$m = 1, \quad \frac{x}{a^5} + \ldots = \frac{1}{p^2}.$$

$$m = 2, \quad \frac{x^2}{a^4} + \ldots = \frac{p^3 \, r_i^3}{p_i^2},$$

$$m = 3, \quad \frac{a^7}{x^7} + \ldots = \frac{p_i^3 \, r_2^3}{p_2^2 \, p_i^4 \, r_i^3}.$$

$$m = 4, \quad \frac{x^{11}}{a^{15}} + \ldots = \frac{p_2^3 \, p_i^7 \, r_3^3 \, r_i^7}{p_3^3 \, p_i^3 \, r_2^2}.$$

$$m = 5, \quad \frac{a^{13}}{x^{13}} + \ldots = \frac{p_3^3 \, p_i^7 \, r_4^3 \, r_2^7}{p_4^2 \, p_2^2 \, p_i^4 \, r_3^3 \, r_i^4}.$$

$$\ldots \qquad \ldots \qquad \ldots$$

(3.)—

$$m = 1, \quad a^{\frac{1}{4}} x^{\frac{1}{4}} + \ldots = p^{\frac{1}{2}} r^{\frac{1}{2}}.$$

$$m = 2, \quad \frac{a^{\frac{2}{3}}}{x^{\frac{2}{3}}} + \ldots = \frac{p^{\frac{1}{2}} \, r_2^{\frac{1}{2}}}{p^{\frac{3}{4}} \, r_i^{\frac{3}{4}}}.$$

$$m = 3, \quad a^{\frac{1}{16}} x^{\frac{7}{16}} + \ldots = \frac{p_2^{\frac{1}{2}} \, n_2^{\frac{2}{3}} \, r_3^{\frac{1}{2}} \, r_i^{\frac{2}{3}}}{p^{\frac{3}{4}} \, r_2^{\frac{2}{4}}}.$$

$$m = 4, \quad \frac{a^{\frac{13}{16}}}{x^{\frac{9}{16}}} + \ldots = \frac{p_3^{\frac{1}{2}} \, p_i^{\frac{3}{8}} \, r_4^{\frac{1}{2}} \, r_2^{\frac{3}{8}}}{p_2^{\frac{1}{4}} \, p^{\frac{3}{16}} \, r_3^{\frac{1}{4}} \, r_i^{\frac{3}{16}}}.$$

$$m = 5, \quad \frac{x^{\frac{29}{32}}}{a^{\frac{23}{32}}} + \ldots = \frac{p_4^{\frac{1}{2}} \, p_2^{\frac{3}{8}} \, p_i^{\frac{11}{32}} \, r_5^{\frac{1}{4}} \, r_3^{\frac{3}{8}} \, r_2^{\frac{31}{32}}}{p_3^{\frac{3}{4}} \, p_i^{\frac{7}{16}} \, r_4^{\frac{1}{4}} \, r_2^{\frac{7}{16}}}.$$

$$\ldots \qquad \ldots \qquad \ldots$$

(XVI.) The instances in which the general formulæ have been here
developed will suffice to elucidate their application to particular surfaces.
At the same time it is evident that it becomes necessary, at this point, to
quit a very interesting, although laborious, field of investigation; since, to
work out in more extensive detail individual illustrations, would be incon-

sistent with the purpose of this book. It may be worth while to call attention to the peculiar symmetry of expression characterising the equations of successive planes in the ellipsoid, as exhibited in page 30; and it is obvious that, in the expanded series, the developments may be rendered in terms of the consecutive values of u, as in the original expressions of page 18, in place of those which have been given in terms of p_m and r_m. Further, it is impossible to overlook the singular relation borne by the symbol ρ_m, in all the surfaces, to the other connected lines.

In conclusion, it will be not uninteresting to examine the law by which the value of the perpendicular p is governed in consecutive surfaces. It is deserving of remark, as indicative of singularity, although it might possibly have been anticipated, that the equation expressing the value of this quantity, as well as the equation of the surface itself, ought to be regarded only as a particular case selected out of an infinite number of connected, symbolised, geometrical relations.

If the reflection seems to be unavoidable, that the co-ordinate relation which it has been customary to call, by long-established convention, " the general equation " of a surface, or a curve, is more properly to be regarded in this light, as exhibiting one only of the aspects under which its subject may be considered; and, in truth, as but a single term in an infinite series, each implying, individually, the essential properties of the figure, an additional reason appears to be presented for limiting, at present, our expectation of geometrical or analytical perfectibility. Considerations of this nature, no doubt, must tend to confirm the impression, which is forced upon our attention from many independent sources, that a far more perfect theory than we as yet possess of symbolised geometry, may be reserved for the research of the future.*

It is, then, to be desired, that those who take interest in abstract speculations upon subjects hitherto but little, if at all, examined, should not shrink from diverging out of beaten tracks of thought; should endeavour, indeed, to fix the impression of ideas which seem to be new; to accumulate, classify, and bring them into prominence : in short, to collect and combine those floating threads of mathematical truth which, singly considered, are barely to be traced; but, collectively, may lend a visible and well-defined support to the beautiful network of science.

* See, on this subject, De Morgan, Double Algebra; and Treatise on Related Caustics, page 12, Cor. 2, by the Author.

(XVII.)   Development of the perpendiculars in consecutive surfaces.

It will be seen by a glance at the tabulated expressions which follow, that both the numerators and denominators in the value of the perpendicular are determined by the simple law of a geometric series in which 2 is the common ratio.

(1.) $\quad \Sigma\left(\dfrac{x}{a}\right) \quad = \quad 1 ; \qquad \dfrac{1}{p^2} \quad = \quad \Sigma\left(\dfrac{1}{a^2}\right).$

(2.) $\quad \Sigma\left(\dfrac{x^2}{a^2}\right) \quad = \quad 1 ; \qquad \dfrac{1}{p^2} \quad = \quad \Sigma\left(\dfrac{x^2}{a^4}\right).$

(3.) $\quad \Sigma\left(\dfrac{x^3}{a^3}\right) \quad = \quad 1 ; \qquad \dfrac{1}{p^2} \quad = \quad \Sigma\left(\dfrac{x^4}{a^6}\right).$

(4.) $\quad \Sigma\left(\dfrac{x^4}{a^4}\right) \quad = \quad 1 ; \qquad \dfrac{1}{p^2} \quad = \quad \Sigma\left(\dfrac{x^6}{a^8}\right).$

(5.) $\quad \Sigma\left(\dfrac{x^5}{a^5}\right) \quad = \quad 1 ; \qquad \dfrac{1}{p^2} \quad = \quad \Sigma\left(\dfrac{x^8}{a^{10}}\right).$

(6.) $\quad \Sigma\left(\dfrac{x^6}{a^6}\right) \quad = \quad 1 ; \qquad \dfrac{1}{p^2} \quad = \quad \Sigma\left(\dfrac{x^{10}}{a^{12}}\right).$

(7.) $\quad \Sigma\left(\dfrac{x^7}{a^7}\right) \quad = \quad 1 ; \qquad \dfrac{1}{p^2} \quad = \quad \Sigma\left(\dfrac{x^{12}}{a^{14}}\right).$

(8.) $\quad \Sigma\left(\dfrac{x^8}{a^8}\right) \quad = \quad 1 ; \qquad \dfrac{1}{p^2} \quad = \quad \Sigma\left(\dfrac{x^{14}}{a^{16}}\right).$

(9.) $\quad \Sigma\left(\dfrac{x^9}{a^9}\right) \quad = \quad 1 ; \qquad \dfrac{1}{p^2} \quad = \quad \Sigma\left(\dfrac{x^{16}}{a^{18}}\right).$

(10.) $\quad \Sigma\left(\dfrac{x^{10}}{a^{10}}\right) \quad = \quad 1 ; \qquad \dfrac{1}{p^2} \quad = \quad \Sigma\left(\dfrac{x^{18}}{a^{20}}\right).$

The same law will be observed to regulate the values of the perpendiculars in consecutive surfaces which are expressed under a negative index; as may be seen in the following series.

(1.) $\quad \Sigma\left(\dfrac{a}{x}\right) \quad = \quad 1; \qquad \dfrac{1}{p^2} \quad = \quad \Sigma\left(\dfrac{a^2}{x^4}\right).$

(2.) $\quad \Sigma\left(\dfrac{a^2}{x^2}\right) \quad = \quad 1; \qquad \dfrac{1}{p^2} \quad = \quad \Sigma\left(\dfrac{a^4}{x^6}\right).$

(3.) $\quad \Sigma\left(\dfrac{a^3}{x^3}\right) \quad = \quad 1; \qquad \dfrac{1}{p^2} \quad = \quad \Sigma\left(\dfrac{a^6}{x^5}\right).$

(4.) $\quad \Sigma\left(\dfrac{a^4}{x^4}\right) \quad = \quad 1; \qquad \dfrac{1}{p^2} \quad = \quad \Sigma\left(\dfrac{a^8}{x^{10}}\right).$

(5.) $\quad \Sigma\left(\dfrac{a^5}{x^5}\right) \quad = \quad 1; \qquad \dfrac{1}{p^2} \quad = \quad \Sigma\left(\dfrac{a^{10}}{x^{12}}\right).$

(6.) $\quad \Sigma\left(\dfrac{a^6}{x^6}\right) \quad = \quad 1; \qquad \dfrac{1}{p^2} \quad = \quad \Sigma\left(\dfrac{a^{12}}{x^{11}}\right).$

(7.) $\quad \Sigma\left(\dfrac{a^7}{x^7}\right) \quad = \quad 1; \qquad \dfrac{1}{p^2} \quad = \quad \Sigma\left(\dfrac{a^{14}}{x^{16}}\right).$

(8.) $\quad \Sigma\left(\dfrac{a^8}{x^8}\right) \quad = \quad 1; \qquad \dfrac{1}{p^2} \quad = \quad \Sigma\left(\dfrac{a^{16}}{x^{18}}\right).$

(9.) $\quad \Sigma\left(\dfrac{a^9}{x^9}\right) \quad = \quad 1; \qquad \dfrac{1}{p^2} \quad = \quad \Sigma\left(\dfrac{a^{18}}{x^{20}}\right).$

(10.) $\quad \Sigma\left(\dfrac{a^{10}}{x^{10}}\right) \quad = \quad 1; \qquad \dfrac{1}{p^3} \quad = \quad \Sigma\left(\dfrac{a^{20}}{x^{22}}\right).$

...   .

K

(XVIII.) By writing $n - 1 = m$, in the general formulæ, page 23, they will be transformed into the following expressions.

$$\Sigma \left\{ \frac{x^{\overset{m}{m}}}{a^{\frac{m}{m-1}} (m+1)m - 2} \right\} = \frac{1}{p_m^m\, p_{m-1}^{\frac{m-1}{m}}\, p_{m-2}^{\frac{2}{m}} \ldots p^{\frac{m-1}{m}}\; r_{m-1}^{\frac{m-1}{m}} r_{m-2} \; r_{,}} \qquad (1)$$

$$\Sigma \left\{ \frac{x^{\overset{m}{m}}}{a^{(m+1)\frac{m-1}{m-1}}} \right\}^{2} = \frac{1}{p_{m-1}^{2}\, p_{m-2}^{2m} \ldots p^{2m}\; r_{m-1}^{\frac{m-1}{2m}} r_{m-2}^{2m} \ldots r_{,}^{\frac{m-1}{2m}}} \qquad (2)$$

$$\Sigma \left\{ \frac{x^{\overset{m}{m}}}{a^{\frac{m}{m-1}} (m+1)m - 2} \right\}^{m+1} = \frac{1}{p_{m-1}^{m+1}\, p_{m-2}^{(m+1)m} \ldots p^{(m+1)m}\; r_m^{m+1} r_{m-1}^{(m+1)m} \ldots r_{,}^{\frac{m-1}{(m+1)m}}} \qquad (3)$$

In employing this modification of the formulæ, consecutive values of m may, of course, be taken, but with each variation in value the surface itself changes; its general type being,

$$\Sigma \left(\frac{x}{a} \right)^{m+1} = 1 \,;$$

in all instances in which n is positive: but, if the index is negative, we have $- n = m + 1$, and the series of surfaces will be represented under the form,

$$\Sigma \left(\frac{a}{x} \right)^{m+1} = 1.$$

Consistently with the hypothesis respecting m, which in its first acceptation has been necessarily assumed to be integral, n must be also integral, in the application of the last set of formulæ; if the surface is such that n is fractional they cannot be employed, in any sense in which the properties of these expressions have been hitherto considered. It may, however, be not impossible that a consistent interpretation is assignable even to fractional values of m. At a first view this may appear paradoxical; yet there does not seem to be any sufficient reason, e.g., why such an ex-

pression as $p_{\frac{1}{2}}$, or $r_{\frac{1}{2}}$, in the preceding pages, should not, in an analytical sense, be as much capable of interpretation as $\dfrac{d^{\frac{1}{2}}u}{d x^{\frac{1}{2}}}$. And we cannot but recollect that results have been obtained in the functional analysis which, before they had been established, were not less beyond anticipation.[*]

If $n = 2$, and consequently $m = 1$, the surface being then the ellipsoid, the index of a, as it has been before remarked, will take the indefinite form $\dfrac{0}{0}$; then,

$$\frac{(m+1) \, m^{m} - 2}{m-1} \;=\; (m+1) \, m^{m} \, (1 + \log m) + m^{m} \;=\; 3 ;$$

$$\frac{(m+1) \, (m^{m} - 1)}{m-1} \;=\; (m+1) \, m^{m} \, (1 + \log m) + m^{m} - 1 \;=\; 2 ;$$

although, perhaps, it may seem hardly necessary to reproduce this evaluation, since it has been already included in page 26, under the original form of the index.

[*] See Gregory's General Differentiation. &c.

CHAPTER IV.

(I.) The object of this chapter is to deduce, in some instances, equations of relation between consecutive radii and perpendiculars in the ellipsoid. It is clear that these relations may exist in infinite variety, and we have to select a few only of those which appear to present the most interesting features in point of form.

Taking the second of the formulæ given at page 26,

$$\Sigma \left\{ \frac{x^2}{a^{\text{im}}} \right\} = \frac{1}{\overset{2}{p_{m-1}} \overset{2}{p_{m-2}} \ldots \overset{2}{p} \ \overset{2}{r_{m-1}} \overset{2}{r_{m-2}} \ldots \overset{2}{r_{\prime}}}.$$

Now, let it be assumed that

$$a_m = \frac{1}{a^{\text{im}}}; \qquad \beta_m = \frac{1}{b^{\text{im}}}; \qquad \gamma_m = \frac{1}{c^{\text{im}}};$$

$$\overset{2}{P}_{m-1} = \frac{1}{\overset{2}{p_{m-1}} \ldots \overset{2}{p} \ \overset{2}{r_{m-1}} \ldots \overset{2}{r_{\prime}}}.$$

then, if we take four consecutive equations, they may be written,

$$a_m \, x^2 + \beta_m \, y^2 + \gamma_m \, z^2 = P^2_{m-1}. \tag{1}$$

$$a_{m-1} \, x^2 + \beta_{m-1} \, y^2 + \gamma_{m-1} \, z^2 = P^2_{m-2}. \tag{2}$$

$$a_{m-2} \, x^2 + \beta_{m-2} \, y^2 + \gamma_{m-2} \, z^2 = P^2_{m-3}. \tag{3}$$

$$a_{m-3} \, x^2 + \beta_{m-3} \, y^2 + \gamma_{m-3} \, z^2 = P^2_{m-4}. \tag{4}$$

From any three of these, when $x^2 \, y^2 \, z^2$ have been determined in terms of the other quantities, their elimination from the remaining equation will lead to a relation between the radii and tangent-perpendiculars of the surface.

As the conclusions which follow may be verified by any one who takes an interest in the subject, it does not seem necessary to express in detail the determination of the co-ordinates. The method which appears to be the most convenient is that of indeterminate factors, keeping in view that the constants a_m β_m γ_m are subject to the law of indices. After the requisite reductions have been made, there will then be, from the three first equations,

$$x^2 = \frac{P^2_{m-1} - (\beta + \gamma) P^2_{m-2} + \beta\gamma P^2_{m-3}}{a_{m-2} (a - \beta) (a - \gamma)}.$$

$$y^2 = -\frac{P^2_{m-1} - (a + \gamma) P^2_{m-2} + a\gamma P^2_{m-3}}{\beta_{m-2} (a - \beta) (\beta - \gamma)}.$$

$$z^2 = \frac{P^2_{m-1} - (a + \beta) P^2_{m-2} + a\beta P^2_{m-3}}{\gamma_{m-2} (a - \gamma) (\beta - \gamma)}.$$

(II.) Of the infinite number of equations derivable from the ellipsoid which may be taken for the elimination of these co-ordinates, we select, in the first instance, that which is next in order of sequence; equation (4), page 68 :

Consequently,

$$\frac{a_{m-3}}{a_{m-2}} \frac{P^2_{m-1} - (\beta + \gamma) P^2_{m-2} + \beta\gamma P^2_{m-3}}{(a - \beta) (a - \gamma)} \quad -$$

$$\frac{\beta_{m-3}}{\beta_{m-2}} \frac{P^2_{m-1} - (a + \gamma) P^2_{m-2} + a\gamma P^2_{m-3}}{(a - \beta) (\beta - \gamma)} \quad +$$

$$\frac{\gamma_{m-3}}{\gamma_{m-3}} \frac{P^2_{m-1} - (a + \beta) P^2_{m-2} + a\beta P^2_{m-3}}{(a - \gamma) (\beta - \gamma)} \quad = \quad P^2_{m-4}.$$

Now, since the suffixed symbols of a β γ follow, in this notation, the law of indices,

$$\frac{a_{m-3}}{a_{m-2}} = \frac{1}{a} \; ; \quad \frac{\beta_{m-3}}{\beta_{m-2}} = \frac{1}{\beta} \; ; \quad \frac{\gamma_{m-3}}{\gamma_{m-2}} = \frac{1}{\gamma} \; ;$$

Let $A_{\prime} = \dfrac{1}{\alpha\,(\alpha-\beta)\,(\alpha-\gamma)} - \dfrac{1}{\beta\,(\alpha-\beta)\,(\beta-\gamma)} + \dfrac{1}{\gamma\,(\alpha-\gamma)\,(\beta-\gamma)}.$

$B_{\prime} = \dfrac{\beta+\gamma}{\alpha\,(\alpha-\beta)\,(\alpha-\gamma)} - \dfrac{\alpha+\gamma}{\beta\,(\alpha-\beta)\,(\beta-\gamma)} + \dfrac{\alpha+\beta}{\gamma\,(\alpha-\gamma)\,(\beta-\gamma)}.$

$C_{\prime} = \dfrac{\beta\gamma}{\alpha\,(\alpha-\beta)\,(\alpha-\gamma)} - \dfrac{\alpha\gamma}{\beta\,(\alpha-\beta)\,(\beta-\gamma)} + \dfrac{\alpha\beta}{\gamma\,(\alpha-\gamma)\,(\beta-\gamma)}.$

The equation which has been deduced may then be written,

$$A_{\prime}\,P^2_{m-1} \; - \; B_{\prime}\,P^2_{m-2} \; + \; C_{\prime}\,P^2_{m-3} \; = \; P^2_{m-4}.$$

It may be shewn, after the requisite reduction, that

$$A_{\prime} \;=\; \frac{1}{\alpha\beta\gamma} \;=\; a^4\,b^4\,c^4$$

$$B_{\prime} \;=\; \Sigma\left\{\frac{1}{\alpha\beta}\right\} \;=\; a^4 b^4 + a^4 c^4 + b^4 c^4.$$

$$C_{\prime} \;=\; \Sigma\left\{\frac{1}{\alpha}\right\} \;=\; a^4 + b^4 + c^4.$$

After increasing by unity the value of m, which is entirely arbitrary, the resulting equation is finally,

$$p^2_m\;p^2_{m-1}\;p^2_{m-2}\;r^2_m\;r^2_{m-1}\;r^2_{m-2} \;\; -$$

$$(a^4 + b^4 + c^4)\,p^2_m\,p^2_{m-1}\,r^2_m\,r^2_{m-1} \;\; +$$

$$(a^4 b^4 + a^4 c^4 + b^4 c^4)\,p_m^2\,r_m^2 \;\; - \;\; a^4 b^4 c^4 = 0. \quad (1.)$$

This equation expresses the relation between any six consecutive perpendiculars and radii from $p_{\prime}$ and $r_{\prime}$ upwards; and, as regards its construction, bears a curious formal analogy to the equation (18) in terms of $r\,p\,r_{\prime}$, deduced in page 4.

A verification of these formulæ may be found when the surface becomes a sphere, in which case all the lines are equal to the radius; in the present case we have, under these circumstances,

$$(r^4 - a^4)^3 \;=\; 0$$

$$\therefore \quad r \;=\; a.$$

(III.) In place of the equation (4), page 68, we now select for the elimination of the co-ordinates the next in succession; viz.—

$$a_{m-4}\, x^2 + \beta_{m-4}\, y^2 + \gamma_{m-4}\, z^2 = P^2_{m-5}.$$

We have, then, $\quad \dfrac{a_{m-4}}{a_{m-2}} = \dfrac{1}{a_2}; \qquad \ldots$

$$A_2 = \frac{1}{a_2\,(a-\beta)\,(a-\gamma)} - \frac{1}{\beta_2(a-\beta)\,\overline{\beta-\gamma}} + \frac{1}{\gamma_2(a-\gamma)\,(\beta-\gamma)}.$$

$$= \frac{1}{a\,\beta\,\gamma}\left(\frac{1}{a} + \frac{1}{\beta} + \frac{1}{\gamma}\right).$$

$$= a^4\,b^4\,c^4\,(a^4 + b^4 + c^4).$$

$$B_2 = \frac{\beta+\gamma}{a_2\,(a-\beta)\,(a-\gamma)} - \frac{a+\gamma}{\beta_2\,(a-\beta)\,(\beta-\gamma)} + \frac{a+\beta}{\gamma_2\,(a-\gamma)\,(\beta-\gamma)}.$$

$$= \left(\frac{1}{a} + \frac{1}{\beta}\right)\left(\frac{1}{a} + \frac{1}{\gamma}\right)\left(\frac{1}{\beta} + \frac{1}{\gamma}\right).$$

$$= (a^4 + b^4)\,(a^4 + c^4)\,(b^4 + c^4).$$

$$C_2 = \frac{\beta\gamma}{a_2\,(a-\beta)\,(a-\gamma)} - \frac{a\gamma}{\beta_2\,(a-\beta)\,(\beta-\gamma)} + \frac{a\beta}{\gamma_2\,(a-\gamma)\,(\beta-\gamma)}.$$

$$= \frac{1}{a_2\,\beta_2\,\gamma_2}\,(a_2\,\beta_2 + a_2\,\gamma_2 + \beta_2\,\gamma_2 + a_2\,\beta\gamma + a\beta_2\,\gamma + a\beta\gamma_2).$$

$$= a^8 + b^8 + c^8 + a^4\,b^4 + a^4\,c^4 + b^4\,c^4.$$

$$= a^4\,(a^4 + b^4) + b^4\,(b^4 + c^4) + c^4\,(c^4 + a^4).$$

By substituting the values of the co-efficients in the equation,

$$A_2\,P^2_{m-1} + B_2\,P^2_{m-2} + C_2\,P^2_{m-3} = P^2_{m-5},$$

we obtain, after increasing the suffix m by unity,

$$p^2_m\, p^2_{m-1}\, p^2_{m-2}\, p^2_{m-3}\, r^2_m\, r^2_{m-1}\, r^2_{m-2}\, r^2_{m-3} -$$

$$\left\{a^4\,(a^4+b^4) + b^4(b^4+c^4) + c^4(c^4+a^4)\right\}\, p^2_m\, p^2_{m-1}\, r^2_m\, r^2_{m-1} +$$

$$(a^4+b^4)\,(a^4+c^4)\,(b^4+c^4)\, p^2_m\, r^2_m - a^4\,b^4\,c^4\,(a^4 + b^4 + c^4) = 0. \quad (2.)$$

This singular equation expresses a relation which obtains between any eight consecutive radii and perpendiculars, four of each, belonging to the ellipsoid. The peculiar symmetry of expression in this, as well as in the preceding equation (1), is deserving of notice ; but, in those of a higher order which will be given subsequently, it is not so much observable : although, still, in all the equations which may be derived from any of these eliminations, the mutual connexion between the lines here subjected to examinaton, no less than their connexion with combinations among the axes of the surface, cannot but be considered as very remarkable.

On reducing the surface to a sphere we find, in this case, as a formula of verification,

$$(r^4 - a^4)\,(r^4 + 3a^4) \;=\; 0$$

$$\therefore \quad r \;=\; a.$$

(IV.) If now we take for the eliminating equation the next in order,

$$\alpha_{m-5}\,x^2 \;+\; \beta_{m-5}\,y^2 \;+\; \gamma_{m-5}\,z^2 \;=\; P^2_{m-6}\;;$$

there will be obtained, after the necessary reductions,

$$A_3 \;=\; a^4\,b^4\,c^4\left\{ a^4\,(a^4 + b^4) + b^4\,(b^4 + c^4) + c^4\,(c^4 + a^4)\right\}.$$

$$B_3 \;=\; a^{12}\,b^4 + a^{12}\,c^4 + b^{12}\,c^4 \;+$$

$$a^8\,b^8 + a^8\,c^8 + b^8\,c^8 \;+$$

$$2a^8\,b^4\,c^4 + 2a^4\,b^8\,c^4 + 2a^4\,b^4\,c^8 \;+$$

$$a^4\,b^{12} + a^4\,c^{12} + b^4\,c^{12}.$$

$$C_3 \;=\; (a^4 + b^4 + c^4)\,(a^8 + b^8 + c^8) + a^4\,b^4\,c^4.$$

Eliminating these co-efficients from the equation,

$$A_3\,P^2_{m-1} \;-\; B_3\,P^2_{m-2} \;+\; C_3\,P^2_{m-3} \;=\; P^2_{m-6},$$

and increasing the value of m by unity, there will be found as the relation between any ten consecutive radii and perpendiculars, five of each, the resulting equation of condition,

$$p^2{}_{m}\ p^2{}_{m-1}\ p^2{}_{m-2}\ p^2{}_{m-3}\ p^2{}_{m-4}\ r^2{}_{m}\ r^2{}_{m-1}\ r^2{}_{m-2}\ r^2{}_{m-3}\ r^2{}_{m-4}\quad -$$

$$\left\{(a^4+b^4+c^4)\,(a^8+b^8+c^8)+a^4\,b^4\,c^4\right\}\,p^2{}_{m}\ p^2{}_{m-1}\ r^2{}_{m}\ r^2{}_{m-1}\quad +$$

$$\left\{a^{13}\,b^4+a^{13}\,c^4+b^{13}\,c^4+a^8\,b^8+a^8\,c^8+b^8\,c^8+2a^8\,b^4\,c^4+2a^4\,b^8\,c^4+2a^4\,b^4\,c^8\right.$$

$$\left.+\ a^4\,b^{13}\ +\ a^4\,c^{13}\ +\ b^4\,c^{13}\right\}\,p^2{}_{m}\ r^2{}_{m}\quad -$$

$$a^4\,b^4\,c^4\left\{a^4\,(a^4+b^4)\ +\ b^4\,(b^4+c^4)\ +\ c^4\,(c^4+a^4)\right\}\ =\ 0.\qquad(3.)$$

On reducing the surface to a sphere we have the formula of verification,

$$r^{20}\ -\ 10\,a^{13}\,r^8\ +\ 15\,a^{16}\,r^4\ -\ 6\,a^{20}\ =\ 0\ ;$$

which is equivalent to the factorial expression,

$$(r^4\ -\ a^4)^3\,(r^8\ +\ 3a^4\,r^4\ +\ 6a^8)\ =\ 0\ :$$

$$\therefore\ \ r\ =\ a.$$

(V.) It is not necessary to adduce further, in this place, individual instances of elimination similar to those which have been here developed; although it may be worth notice, that, as they are capable of endless extension, it is not unlikely that many very curious combinations might be elicited. Instead, then, of dwelling longer, at present, upon particular cases, we shall examine the general forms of the co-efficients.

The general form of the eliminating equation is,

$$a_{m-k}\,x^2\ +\ \beta_{m-k}\,y^2\ +\ \gamma_{m-k}\,z^2\ =\ \mathrm{P}^2{}_{m-k-1}\ ;$$

and it is evident that, as the value of k becomes greater, the complexity of the final equation, as well as the number of lines which it involves, will be increased.

L

Let the corresponding co-efficients be written,

$$A_{k-2} ; \qquad B_{k-2} ; \qquad C_{k-2} ;$$

then,
$$\frac{a_{m-k}}{a_{m-2}} = \frac{1}{a_{k-2}} ; \qquad \&c. :$$

and we shall have the following expressions,

$$A_{k-2} = \frac{1}{a_{k-2}\,(a-\beta)\,(a-\gamma)} - \frac{1}{\beta_{k-2}\,(a-\beta)\,\beta-\gamma)} + \frac{1}{\gamma_{k-2}\,(a-\gamma)\,(\beta-\gamma)}.$$

$$B_{k-2} = \frac{\beta+\gamma}{a_{k-2}\,(a-\beta)\,(a-\gamma)} - \frac{a+\gamma}{\beta_{k-2}\,(a-\beta)\,(\beta-\gamma)} + \frac{a+\beta}{\gamma_{k-2}\,(a-\gamma)\,(\beta-\gamma)}.$$

$$C_{k-2} = \frac{\beta\gamma}{a_{k-2}\,(a-\beta)\,(a-\gamma)} - \frac{a\gamma}{\beta_{k-2}\,(a-\beta)\,(\beta-\gamma)} + \frac{a\beta}{\gamma_{k-2}\,(a-\gamma)\,(\beta-\gamma)}.$$

consequently,

$$A_{k-2} = \frac{a_{k-2}\,\beta_{k-2}\,(a-\beta) - a_{k-2}\,\gamma_{k-2}\,(a-\gamma) + \beta_{k-2}\,\gamma_{k-2}\,(\beta-\gamma)}{a_{k-2}\,\beta_{k-2}\,\gamma_{k-2}\,(a-\beta)\,(a-\gamma)\,(\beta-\gamma)}.$$

$$B_{k-2} = \frac{a_{k-2}\,\beta_{k-2}(a_2-\beta_2) - a_{k-2}\,\gamma_{k-2}(a_2-\gamma_2) + \beta_{k-2}\,\gamma_{k-2}(\beta_2-\gamma_2)}{a_{k-2}\,\beta_{k-3}\,\gamma_{k-2}\,(a-\beta)\,(a-\gamma)\,(\beta-\gamma)}.$$

$$C_{k-2} = \frac{a_{k-1}\,\beta_{k-1}\,(a-\beta) - a_{k-1}\,\gamma_{k-1}\,(a-\gamma) + \beta_{k-1}\,\gamma_{k-1}\,(\beta-\gamma)}{a_{k-2}\,\beta_{k-2}\,\gamma_{k-2}\,(a-\beta)\,(a-\gamma)\,(\beta-\gamma)}.$$

In all cases, the numerators of these fractions are divisible without remainder by the binomial factors in the denominators; this is evident, because either of the assumptions, $a - \beta = 0$; $a - \gamma = 0$; $\beta - \gamma = 0$; will cause both of the terms to vanish : and it would not be difficult to expand them into series, according to the method afterwards

adopted in regard to similar functions. There is, however, no necessity to exhibit these quantities under such a form, since, for adaptation to particular cases, the present arrangement is sufficiently convenient.

(VI) After introducing the values of the constants which have been now determined, into the general equation, viz.:

$$A_{k-2} P^2_{m-1} - B_{k-2} P^2_{m-2} + C_{k-2} P^2_{m-3} = P^2_{m-k-1} ;$$

the general equation will result,

$$p^2_{m-1} \cdots p^2_{m-k} r^2_{m-1} \cdots r^2_{m-k} -$$

$$C_{k-2} \, p^2_{m-1} \, p^2_{m-2} \, r^2_{m-1} \, r^2_{m-2} \; +$$

$$B_{k-2} \, p^2_{m-1} \, r^2_{m-1} \; -$$

$$A_{k-2} = 0.$$

(VII.) The line r, the first vectorial radius, does not appear in any of the preceding equations, after (13), page 4; it may be introduced by selecting for the eliminating equation,

$$x^2 + y^2 + z^2 = r^2.$$

Then, from page 69,

$$\frac{P^2_{m-1} - (\beta + \gamma) P^2_{m-2} + \beta\gamma P^2_{m-3}}{a_{m-2} (a - \beta) (a - \gamma)} -$$

$$\frac{P^2_{m-1} - (a + \gamma) P^2_{m-2} + a\gamma P^2_{m-3}}{\beta_{m-2} (a - \beta) (\beta - \gamma)} +$$

$$\frac{P^2_{m-1} - (a + \beta) P^2_{m-2} + a\beta P^2_{m-3}}{\gamma_{m-2} (a - \gamma) (\beta - \gamma)} = r^2.$$

(VIII.) We proceed to develop into series the co-efficients involved in the preceding elimination; the first of which will be,

$$A_{m-2} = \frac{a_{m-2}\,\beta_{m-2}\,(a-\beta) - a_{m-2}\,\gamma_{m-2}\,(a-\gamma) + \beta_{m-2}\,\gamma_{m-2}\,(\beta-\gamma)}{a_{m-2}\,\beta_{m-2}\,\gamma_{m-2}\,(a-\beta)\,(a-\gamma)\,(\beta-\gamma)}.$$

Let the denominator be represented, at first, by D, and by D, D_2 when the fraction has been cleared of the factors $(\beta-\gamma)$ and $(a-\gamma)$; then,

$$A_{m-2} = \frac{1}{D}\left\{ a_{m-1}\,(\beta_{m-2}-\gamma_{m-2}) - a_{m-2}(\beta_{m-1}-\gamma_{m-1}) + \beta_{m-2}\,\gamma_{m-2}\,(\beta-\gamma) \right\}.$$

$$= \frac{1}{D_{\prime}}\left\{ a_{m-1}\,(\beta_{m-3} + \beta_{m-4}\,\gamma + \ldots + \beta\gamma_{m-4} + \gamma_{m-3}) \right.$$

$$- \; a_{m-2}\,(\beta_{m-2} + \beta_{m-3}\,\gamma + \ldots + \beta\gamma_{m-3} + \gamma_{m-2})$$

$$\left. + \; \beta_{m-2}\,\gamma_{m-2} \right\}.$$

Since the expansion of $\dfrac{\beta_k - \gamma_k}{\beta - \gamma}$ contains k terms, there will be here $(m-2) + (m-1) + 1$ terms, $= 2(m-1)$; and, since this is an even number, they may be combined in $(m-1)$ terms of a binominal form; then, after two reductions,

$$A_{m-2} = \frac{1}{D_{\prime}}\left\{ (a-\gamma)\,a_{m-2}\,(\beta_{m-3} + \beta_{m-4}\,\gamma + \ldots + \beta\gamma_{m-4}+\gamma_{m-3}). \right.$$

$$\left. - \; \beta_{m-2}\,(a_{m-2} - \gamma_{m-2}) \right\}.$$

$$= \frac{1}{D_{2}}\left\{ a_{m-2}\,(\beta_{m-3} + \beta_{m-4}\,\gamma + \ldots + \beta\gamma_{m-4} + \gamma_{m-3}) \right.$$

$$\left. - \; \beta_{m-2}\,(a_{m-3} + a_{m-4}\,\gamma + \ldots + a\gamma_{m-4} + \gamma_{m-3}) \right\}.$$

The first of the two component series involves $m - 2$ terms, and the second an equal number; in all, therefore, there are $2(m - 2)$ terms, which may be incorporated into $m - 2$ binomials, each of which includes the factor $(a - \beta)$: after the expansions have been effected we shall find, ultimately,

$$A_{m-2} = \frac{1}{a_{m-2}\ \beta_{m-2}\ \gamma_{m-2}} \left\{ a_{m-3}\ \beta_{m-3} + a_{m-4}\ \beta_{m-4}(a+\beta)\gamma \right.$$

$$+\ a_{m-5}\ \beta_{m-5}\ (a_2 + a\beta + \beta_2)\ \gamma_2$$

$$+\ \dots$$

This series will always end with the first term involving $a_0\ \beta_0$, a factor which is equal to unity.

The number of terms included in this co-efficient is

$$=\ 1 + 2 + 3 + \dots\ \text{to}\ m - 2\ \text{terms.}$$

$$=\ \frac{1}{2}\ (m - 1)\ (m - 2).$$

For the determination of the second co-efficient we have,

$$B_{m-2} = \frac{1}{D}\left\{ a_{0i}(\beta_{m-2} - \gamma_{m-2}) - a_{m-2}(\beta_m - \gamma_m) + \beta_{m-2}\gamma_{m-2}(\beta_2 - \gamma_2) \right\}.$$

$$=\ \frac{1}{D_i}\left\{ a_{1n}\ (\beta_{m-3} + \beta_{m-4}\ \gamma + \dots\ \text{to}\ m-2\ \text{terms}) \right.$$

$$-\ a_{m-2}\ (\beta_{m-1} - \beta_{m-2}\ \gamma + \dots\ \text{to}\ m\ \text{terms})$$

$$+\ \beta_{m-2}\ \gamma_{m-2}\ (\beta + \gamma)$$

In this expansion there are, in all, $(m-2) + (m+2)$ terms, $= 2m$, and we remove the two last from the second line, combining them with the third line; then,

$$B_{m-2} = \frac{1}{D_{\prime}} \left\{ a_m (\beta_{m-3} + \beta_{m-4}\,\gamma + \ldots \quad \text{to } m-2 \text{ terms}) \right.$$

$$- \quad a_{m-2} (\beta_{m-1} + \beta_{m-2}\,\gamma + \ldots \quad \text{to } m-2 \text{ terms})$$

$$- \quad (a_{m-2} - \beta_{m-2})\,\beta\gamma_{m-2} - (a_{m-2} - \beta_{m-2})\,\gamma_{m-1}$$

$$= \frac{1}{D_{\prime}} \left\{ a_{m-2} (a_2 - \beta_2)(\beta_{m-3} + \beta_{m-4}\,\gamma + \ldots) \right.$$

$$\left. - \quad \beta\gamma_{m-2} (a_{m-2} - \beta_{m-2}) - \gamma_{m-1} (a_{m-2} - \beta_{m-2}) \right\}$$

In this arrangement there are m binomials, and in the next expansion we divide by $(a-\beta)$, calling

$$D_2 = a_{m-2}\,\beta_{m-2}\,\gamma_{m-2}\,(a-\gamma) ;$$

$$\therefore \; B_{m-2} = \frac{1}{D_2} \left\{ a_{m-2} (a+\beta)(\beta_{m-3} + \beta_{m-4}\,\gamma + \ldots) \right.$$

$$- \quad (a_{m-3} + a_{m-4}\,\beta + \ldots)\,\beta\gamma_{m-2}$$

$$\left. - \quad (a_{m-3} + a_{m-4}\,\beta + \ldots)\,\gamma_{m-1} \right\}$$

This expression contains $4(m-2)$ single terms, which may be incorporated into $2(m-2)$ binomials, each involving the factor $(a-\gamma)$. In order to this the two last serial terms must be inverted, and we shall have,

$$B_{m-2} = \frac{1}{D_2} \left\{ a_{m-1}\,\beta_{m-3} + a_{m-2}\,\beta_{m-2} + a_{m-1}\,\beta_{m-4}\,\gamma + a_{m-2}\,\beta_{m-3}\gamma \right.$$

$$+ \quad a_{m-1}\,\beta_{m-5}\,\gamma_2 + a_{m-2}\,\beta_{m-4}\,\gamma_2 + \ldots$$

$$- \quad (\beta_{m-3} + a\beta_{m-4} + \ldots)\,\beta\gamma_{m-2}$$

$$\left. - \quad (\beta_{m-3} + a\beta_{m-4} + \ldots)\,\gamma_{m-1} \right.$$

$$= \frac{1}{D_2} \Big\{ (a_{m-1} - \gamma_{m-1}) \beta_{m-3} + (a_{m-2} - \gamma_{m-2}) a\beta_{m-4}\gamma$$

$$+ \; ... \; \text{to } m - 2 \text{ terms.}$$

$$+ \; (a_{m-2} - \gamma_{m-2}) \beta_{m-2} + (a_{m-3} - \gamma_{m-3}) a\beta_{m-3}\gamma$$

$$+ \; ... \; \text{to } m - 2 \text{ terms.}$$

$$B_{m-2} = \frac{1}{a_{m-2}\,\beta_{m-2}\,\gamma_{m-2}} \Big\{ (a_{m-2}+a_{m-3}\,\gamma + \; ...\;) \beta_{m-3}$$

$$+ \; (a_{m-3} + a_{m-4}\,\gamma + \; ...\;) a\beta_{m-4}\,\gamma$$

$$+ \; ...$$

$$+ \; (a_{m-3} + a_{m-4}\,\gamma + \; ...\;) \beta_{m-2}$$

$$+ \; (a_{m-4} + a_{m-5}\,\gamma + \; ...\;) a\beta_{m-3}\,\gamma$$

$$+ \; ...$$

In each of the two sets composing this co-efficient there are $(m - 2)$ series, the terms of which form arithmetic progressions in regard to the suffixed symbols, with the common difference $(- 1)$; hence,

in 1st set, sum of terms $= (m-1) + (m-2) + ... = \frac{1}{2}(m-2)(m+1)$:

,, 2nd ,, ,, ,, $= (m-2) + (m-3) + ... = \frac{1}{2}(m-2)(m-1)$:

consequently the whole number of terms in B_{m-2} is $= m(m-2)$.

The third co-efficient will be,

$$C_{m-3} = \frac{1}{D}\left\{ a_m(\beta_{m-1}-\gamma_{m-1}) - a_{m-1}(\beta_m-\gamma_m) + \beta_{m-1}\gamma_{m-1}(\beta-\gamma)\right\}.$$

$$= \frac{1}{D_{,}}\left\{ a_m(\beta_{m-2}+\beta_{m-3}\gamma+\dots)\right.$$

$$- a_{m-1}(\beta_{m-1}+\beta_{m-2}\gamma+\dots)$$

$$+ \beta_{m-1}\gamma_{m-1}.$$

This expression contains $(m-1) + (m+1)$ terms, $= 2m$, which may be incorporated into m binomials; then we find,

$$C_{m-2} = \frac{1}{D_{,}}\left\{ a_{m-1}\beta_{m-2}(a-\gamma)\right.$$

$$+ a_{m-1}\beta_{m-3}\gamma(a-\gamma)$$

$$+$$

$$- \beta_{m-1}(a_{m-1}-\gamma_{m-1}).$$

$$= \frac{1}{D_2}\left\{ a_{m-1}\beta_{m-2}+a_{m-1}\beta_{m-3}\gamma+\dots\right.$$

$$- \beta_{m-1}(a_{m-2}+a_{m-3}\gamma+\dots).$$

The last series may be resolved into $(m-1)$ binomials, each containing the factor $(a-\beta)$; hence there is, finally,

$$C_{m-2} = \frac{1}{a_{m-2}\beta_{m-2}\gamma_{m-2}}\left\{ a_{m-2}\beta_{m-2}+a_{m-3}\beta_{m-3}(a+\beta)\gamma\right.$$

$$+ a_{m-4}\beta_{m-4}(a^2+a\beta+\beta^2)\gamma^2$$

$$+ \dots$$

(IX.) The suffixed symbol m may be increased by unity throughout the investigation; and, before extending their application to particular cases, it is convenient to recapitulate the expansions which have been deduced for the several co-efficients: in doing this, since the quantities which enter into the series are similarly involved, the symbols $\beta\gamma$, in the development of B_{m-1}, may be transposed, so that we shall have,

$$A_{m-1} = \frac{1}{a_{m-1}\,\beta_{m-1}\,\gamma_{m-1}} \left\{ a_{m-2}\,\beta_{m-2} + a_{m-3}\,\beta_{m-3}\,(a+\beta)\,\gamma \right.$$
$$+ \quad a_{m-4}\,\beta_{m-4}\,(a_2 + a\beta + \beta_2)\,\gamma_2$$
$$+ \quad \dots$$

$$B_{m-1} = \frac{1}{a_{m-1}\,\beta_{m-1}\,\gamma_{m-1}} \left\{ (a_{m-1} + a_{m-2}\,\beta + \dots)\,\gamma_{m-2} \right.$$
$$+ \quad (a_{m-2} + a_{m-3}\,\beta + \dots)\,a\beta\gamma_{m-3}$$
$$+ \quad (a_{m-3} + a_{m-4}\,\beta + \dots)\,a_2\,\beta_2\,\gamma_{m-4}$$
$$+ \quad \dots$$
$$+ \quad (a_{m-2} + a_{m-3}\,\beta + \dots)\,\gamma_{m-1}$$
$$+ \quad (a_{m-3} + a_{m-4}\,\beta + \dots)\,a\beta\gamma_{m-2}$$
$$+ \quad (a_{m-4} + a_{m-5}\,\beta + \dots)\,a_2\,\beta_2\,\gamma_{m-3}$$
$$+ \quad \dots$$

$$C_{m-1} = \frac{1}{a_{m-1}\,\beta_{m-1}\,\gamma_{m-1}} \left\{ a_{m-1}\,\beta_{m-1} + a_{m-2}\,\beta_{m-2}\,(a+\beta)\,\gamma \right.$$
$$+ \quad a_{m-3}\,\beta_{m-3}\,(a_2 + a\beta + \beta_2)\,\gamma_2$$
$$+ \quad \dots$$

M

In these expressions, A_{m-1} contains $\dfrac{m(m-1)}{2}$ terms

$$B_{m-1} \quad,, \quad (m^2-1) \quad,,$$

$$C_{m-1} \quad,, \quad \dfrac{m(m+1)}{2} \quad,,$$

Cor. It is evident that the co-efficient C_{m-1} might have been inferred, without an independent development, by writing $m+1$ for m in the numerator of A_{m-1}.

(X.) After m has been augmented by unity in (VII), page 75, and a proper arrangement given to the terms, there will be found,

$$r^2 = A_{m-1}\, P^2_m - B_{m-1}\, P^2_{m-1} + C_{m-1}\, P^2_{m-2}.$$

which is equivalent to the general equation,

$$p^2_m\ p^2_{m-1} \ldots p_i^{\,2}\ p^2\ r^2_m\ r^2_{m-1} \ldots r_i^{\,2}\ r^2$$

$$- \quad C_{m-1}\ p^2_m\ p^2_{m-1}\ r^2_m\ r^2_{m-1}$$

$$+ \quad B_{m-1}\ p^2_m\ r^2_m$$

$$- \quad A_{m-1} \quad = \quad 0.$$

(XI.) Some developments of these formulæ will here be given; in each of them it is to be recollected that the series will end so soon as a term appears containing a_0, β_0, or γ_0.

(1.)—Let $m = 2$.

$$A_{,} = \frac{1}{a\beta\gamma} = a^4\,b^4\,c^4.$$

$$B_{,} = \frac{1}{a\beta\gamma}\Big\{(a+\beta) + \gamma\Big\} = \frac{1}{a\beta}+\frac{1}{a\gamma}+\frac{1}{\beta\gamma} = a^4 b^4 + a^4 c^4 + b^4 c^4.$$

$$C_{,} = \frac{1}{a\beta\gamma}\Big\{a\beta + (a+\beta)\gamma\Big\} = \frac{1}{a} + \frac{1}{\beta} + \frac{1}{\gamma} = a^4 + b^4 + c^4.$$

The consequent relation between the six lines r r_i r_2 p p_i p_2 is, then,

$$p^2_{\,2}\, p_i^{\,2}\, p^2\, r^2_{\,2}\, r_i^{\,2}\, r^2 - (a^4 + b^4 + c^4)\, p^2_{\,2}\, p_i^{\,2}\, r_2^{\,2}\, r_i^{\,2}$$

$$+\ (a^4 b^4 + a^4 c^4 + b^4 c^4)\, p^2_{\,2}\, r^2_{\,2} - a^4 b^4 c^4 = 0.$$

This equation has been already obtained by an independent elimination in (1), page 70, if we there write $m = 2$; considering $r_0 = r.$ $p_0 = p.$

(2.)—Let $m = 3.$

$$A_2 = \frac{1}{a_2 \beta_2 \gamma_2}\left\{\alpha\beta + \alpha\gamma + \beta\gamma\right\} = a^4 b^4 c^4 (a^4 + b^4 + c^4).$$

$$B_2 = \frac{1}{a_2 \beta_2 \gamma_2}\left\{(a_2 + \alpha\beta + \beta_2)\gamma + (\alpha+\beta)\alpha\beta + (\alpha+\beta)\gamma_2 + \alpha\beta\gamma\right\}.$$

$$=\ (a^4 + b^4)(a^4 + c^4)(b^4 + c^4).$$

$$C_2 = \frac{1}{a_2 \beta_2 \gamma_2}\left\{a_2 \beta_2 + \alpha\beta(\alpha+\beta)\gamma + (a_2 + \alpha\beta + \beta_2)\gamma_2\right\}.$$

$$=\ a^4(a^4 + b^4) + b^4(b^4 + c^4) + c^4(c^4 + a^4).$$

Hence, as in (2), page 71,

$$p^2_{\,3}\, p^2_{\,2}\, p_i^{\,2}\, p^2\, r^2_{\,3}\, r^2_{\,2}\, r_i^{\,2}\, r^2\ -$$

$$\left\{a^4(a^4 + b^4) + b^4(b^4 + c^4) + c^4(c^4 + a^4)\right\} p^2_{\,3}\, p^2_{\,2}\, r^2_{\,3}\, r^2_{\,2}\ +$$

$$(a^4 + b^4)(a^4 + c^4)(b^4 + c^4)\, p^2_{\,3}\, r^2_{\,3} - a^4 b^4 c^4 (a^4 + b^4 + c^4) = 0.$$

(3.)—In the developments which follow, the co-efficients will be exhibited, without any further reduction into factors, under the symmetrical forms which they naturally assume.

Let $m = 4$.

$$A_2 = a^4 b^4 c^4 \left\{ (a^8 + a^4 b^4 + b^8) + (a^4 + b^4) c^4 + c^8 \right\}.$$

$$
\begin{aligned}
B_2 = \;& (a^{12} + a^8 b^4 + a^4 b^8 + b^{12}) c^4 \\
& + (a^8 + a^4 b^4 + b^8) c^8 \\
& + (a^4 + b^4) c^{12} \\
& + (a^8 + a^4 b^4 + b^8) a^4 b^4 \\
& + (a^4 + b^4) a^4 b^4 c^4 \\
& + a^4 b^4 c^8.
\end{aligned}
$$

$$
\begin{aligned}
C_2 = \;& (a^{12} + a^8 b^4 + a^4 b^8 + b^{12}) \\
& + (a^8 + a^4 b^4 + b^8) \, c^4 \\
& + (a^4 + b^4) c^8 \\
& + c^{12}.
\end{aligned}
$$

Hence, the relation between ten consecutive lines is found to be,

$$p^2_4 \, p^2_3 \, p^2_2 \, p_1^2 \, p^2 \, r^2_4 \, r^2_3 \, r^2_2 \, r_1^2 \, r^2 -$$

$$
\begin{aligned}
& \left\{ (a^{12} + a^8 b^4 + a^4 b^8 + b^{12}) \right. \\
& \quad + (a^8 + a^4 b^4 + b^8) c^4 \\
& \qquad + (a^4 + b^4) c^8 + c^{12} \left.\right\} p^2_4 \, p^2_3 \, r^2_4 \, r^2_3 \; + \\
& \left\{ (a^{12} + a^8 b^4 + a^4 b^8 + b^{12}) c^4 \right. \\
& \quad + (a^8 + a^4 b^4 + b^8) c^8 \\
& \qquad + (a^4 + b^4) c^{12} \\
& \quad + (a^8 + a^4 b^4 + b^8) a^4 b^4 \\
& \qquad + (a^4 + b^4) a^4 b^4 c^4 \\
& \qquad\quad + a^4 b^4 c^8 \left.\right\} p^2_4 \, r^2_4 \; - \\
& a^4 b^4 c^4 \left\{ (a^8 + a^4 b^4 + b^8) + (a^4 + b^4) c^4 + c^8 \right\} = 0.
\end{aligned}
$$

In verification of this equation, the reduction to the sphere will give,
$1 - 10 + 15 - 6 = 0$, identically.

(4.)—Let $m = 5$.

$$A_4 = a^4 b^4 c^4 \left\{ (a^{12} + a^8 b^4 + a^4 b^8 + b^{12}) \right.$$
$$+ (a^8 + a^4 b^4 + b^8) c^4$$
$$+ (a^4 + b^4) c^8$$
$$\left. + c^{12} \right\}.$$

$$B_4 = (a^{16} + a^{12} b^4 + a^8 b^8 + a^4 b^{12} + b^{16}) c^4$$
$$+ (a^{12} + a^8 b^4 + a^4 b^8 + b^{12}) c^8$$
$$+ (a^8 + a^4 b^4 + b^8) c^{12}$$
$$+ (a^4 + b^4) c^{16}$$
$$+ (a^{12} + a^8 b^4 + a^4 b^8 + b^{12}) a^4 b^4$$
$$+ (a^8 + a^4 b^4 + b^8) a^4 b^4 c^4$$
$$+ (a^4 + b^4) a^4 b^4 c^8$$
$$+ a^4 b^4 c^{12}$$

$$C_4 = (a^{16} + a^{12} b^4 + a^8 b^8 + a^4 b^{12} + b^{16})$$
$$+ (a^{12} + a^8 b^4 + a^4 b^8 + b^{12}) c^4$$
$$+ (a^8 + a^4 b^4 + a^8) c^8$$
$$+ (a^4 + b^4) c^{12}$$
$$+ c^{16}$$

These values of the co-efficients having been introduced into the general equation, there will be obtained as the relation between the first 12 lines :

$$p^2{}_5 \ \ p^2{}_4 \ \ p^2{}_3 \ \ p^2{}_2 \ \ p_i{}^2 \ \ p^2 \ \ r^2{}_5 \ \ r^2{}_4 \ \ r^2{}_3 \ \ r^2{}_2 \ \ r_i{}^2 \ \ r^2$$

$$- \ \mathrm{C}_4 \ p^2{}_5 \ p^2{}_4 \ r^2{}_5 \ r^2{}_4 \ + \ \mathrm{B}_4 \ p^2{}_5 \ r^2{}_5 \ - \ \mathrm{A}_4 \ = \ 0$$

(5.)—Let $m = 6$.

The relation between 14 lines will then be,

$$p^2{}_6 \ \ p^2{}_5 \ \ p^2{}_4 \ \ p^2{}_3 \ \ p^2{}_2 \ \ p_i{}^2 \ \ p^2 \ \ r^2{}_6 \ \ r^2{}_5 \ \ r^2{}_4 \ \ r^2{}_3 \ \ r^2{}_2 \ \ r_i{}^2 \ \ r^2$$

$$- \ \mathrm{C}_5 \ p^2{}_6 \ p^2{}_5 \ r^2{}_6 \ r^2{}_5 \ + \ \mathrm{B}_5 \ p^2{}_6 \ r^2{}_6 \ - \ \mathrm{A}_5 \ = \ 0.$$

$$\begin{aligned}
\mathrm{A}_5 \ = \ a^4 \, b^4 \, c^4 \ \Big\{ &(a^{16} + a^{12} \, b^4 + a^8 \, b^8 + a^4 \, b^{12} + b^{16}) \\
&+ \ (a^{12} + a^8 \, b^4 + a^4 \, 2^8 + b^{12}) \, c^4 \\
&+ \ (a^8 + a^4 \, b^4 + b^8) \, c^8 \\
&+ \ (a^4 + b^4) \, c^{12} \\
&+ \ c^{16} \Big\}.
\end{aligned}$$

$$\begin{aligned}
\mathrm{B}_5 \ = \ &(a^{20} + a^{16} \, b^4 + a^{12} \, b^8 + a^8 \, b^{12} + a^4 \, b^{16} + b^{20}) \, c^4 \\
&+ \ (a^{16} + a^{12} \, b^4 + a^8 \, b^8 + a^4 \, b^{12} + b^{16}) \, c^8 \\
&+ \ (a^{12} + a^8 \, b^4 + a^4 \, b^8 + b^{12}) \, c^{12} \\
&+ \ (a^8 + a^4 \, b^4 + b^8) \, c^{16} \\
&+ \ (a^4 + b^4) \, c^{20}. \\
&+ \ (a^{16} + a^{12} \, b^4 + a^8 \, b^8 + a^4 \, b^{12} + b^{16}) \, a^4 \, b^4 \\
&+ \ (a^{12} + a^8 \, b^4 + a^4 \, b^8 + b^{12}) \, a^4 \, b^4 \, c^4 \\
&+ \ (a^8 + a^4 \, b^4 + b^8) \, a^4 \, b^4 \, c^8 \\
&+ \ (a^4 + b^4) \, a^4 \, b^4 \, c^{12} \\
&+ \ a^4 \, b^4 \, c^{16}
\end{aligned}$$

$$C_5 = (a^{20} + a^{16} b^4 + a^{12} b^8 + a^8 b^{12} + a^4 b^{16} + b^{20})$$

$$+ (a^{16} + a^{12} b^4 + a^8 b^8 + a^4 b^{12} + b^{16}) c^4$$

$$+ (a^{12} + a^8 b^4 + a^4 b^8 + b^{12}) c^8$$

$$+ (a^8 + a^4 b^4 + b^8) c^{12}$$

$$+ (a^4 + b^4) c^{16}$$

$$+ c^{20}.$$

In A_5 there are $\dfrac{1}{2} m (m-1)$ terms $= \quad 15.$

B_5 „   „ (m^2-1) „ $= \quad 35.$

C_5 „   „ $\dfrac{1}{2} m (m+1)$ „ $= \quad 21.$

As a verification, when the surface is reduced to a sphere, the equation becomes,

$$1 - 21 + 35 - 15 = 0 ; \text{ identically.}$$

(6.)—Let $m = 7.$

The equation between the first 16 radii and perpendiculars is then,

$$p^2{}_7 \ p^2{}_6 \ p^2{}_5 \ p^2{}_4 \ p^2{}_3 \ p^2{}_2 \ p_i{}^2 \ p^2 \ r^2{}_7 \ r^2{}_6 \ r^2{}_5 \ r^2{}_4 \ r^2{}_3 \ r^2{}_2 \ r_i{}^2 \ r^2$$

$$- C_6 \ p^2{}_7 \ p^2{}_6 \ r^2{}_7 \ r^2{}_6 \ + \ B_6 \ p^2{}_7 \ r^2{}_7 \ - \ A_6 \ = \ 0.$$

In which there will be found, for the values of the co-efficients, the following expressions ;

$$A_6 = a^4 b^4 c^4 \left\{ (a^{20} + a^{16} b^4 + a^{12} b^8 + a^8 b^{12} + a^4 b^{16} + b^{20}) \right.$$

$$+ \; a^{16} + a^{12} b^4 + a^8 b^8 + a^4 b^{12} + b^{16}) \, c^4$$

$$+ \; (a^{12} + a^8 b^4 + a^4 b^8 + b^{12}) \, c^8$$

$$+ \; (a^8 + a^4 b^4 + b^8) \, c^{12}$$

$$+ \; (a^4 + b^4) \, c^{16}$$

$$\left. + \; c^{20} \right\} .$$

$$B_6 = (a^{24} + a^{20} b^4 + a^{16} b^8 + a^{12} b^{12} + a^8 b^{16} + a^4 b^{20} + b^{24}) \, c^4$$

$$+ \; (a^{20} + a^{16} b^4 + a^{12} b^8 + a^8 b^{12} + a^4 b^{16} + b^{20}) \, c^8$$

$$+ \; (a^{16} + a^{12} b^4 + a^8 b^8 + a^4 b^{12} + b^{16}) \, c^{12}$$

$$+ \; (a^{12} + a^8 b^4 + a^4 b^8 + b^{12}) \, c^{16}$$

$$+ \; (a^8 + a^4 b^4 + b^8) \, c^{20}$$

$$+ \; (a^4 + b^4) \, c^{24}.$$

$$+ \; (a^{20} + a^{16} b^4 + a^{12} b^8 + a^8 b^{12} + a^4 b^{16} + b^{20}) \, a^4 b^4$$

$$+ \; (a^{16} + a^{12} b^4 + a^8 b^8 + a^4 b^{12} + b^{16}) \, a^4 b^4 c^4$$

$$+ \; (a^{12} + a^8 b^4 + a^4 b^8 + b^{12}) \, a^4 b^4 c^8$$

$$+ \; (a^8 + a^4 b^4 + b^8) \, a^4 b^4 c^{12}$$

$$+ \; (a^4 + b^4) \, a^4 b^4 c^{16}$$

$$+ \; a^4 b^4 c^{20}.$$

$$C_6 = (a^{24} + a^{20} b^4 + a^{16} b^8 + a^{12} b^{12} + a^8 b^{16} + a^4 b^{20} + b^{24})$$

$$+ (a^{20} + a^{16} b^4 + a^{12} b^8 + a^8 b^{12} + a^4 b^{16} + b^{20}) c^4$$

$$+ (a^{16} + a^{12} b^4 + a^8 b^8 + a^4 b^{12} + b^{16}) c^8$$

$$+ (a^{12} + a^8 b^4 + a^4 b^8 + b^{12}) c^{12}$$

$$+ (a^8 + a^4 b^4 + b^8) c^{16}$$

$$+ (a^4 + b^4) c^{20}.$$

$$+ c^{24}.$$

In A_6 there are here, $\frac{1}{2} m (m-1)$ terms $= 21.$

B_6 „ „ (m^2-1) „ $= 48.$

C_6 „ „ $\frac{1}{2} m (m+1)$ „ $= 28.$

The reduction to the sphere will give the identical equation,

$$1 - 28 + 48 - 21 = 0.$$

(6.)—Let $m = 8.$

The equation between the first 18 radii and perpendiculars will be,

$$p^2{}_8 \; p^2{}_7 \; p^2{}_6 \; p^2{}_5 \; p^2{}_4 \; p^2{}_3 \; p^2{}_2 \; p_l^2 \; p^2 \; r^2{}_8 \; r^2{}_7 \; r^2{}_6 \; r^2{}_5 \; r^2{}_4 \; r^2{}_3 \; r^2{}_2 \; r_l^2 \; r^3$$

$$- C_7 \; p^2{}_8 \; p^2{}_7 \; r^2{}_8 \; r^2{}_7 \; + \; B_7 \; p^2{}_8 \; r^2{}_8 \; - \; A_7 \; = \; 0.$$

$$A_7 = a^4\, b^4\, c^4\, \Big\{ (a^{24} + a^{20}\, b^4 + a^{16}\, b^8 + a^{12}\, b^{12} + a^8\, b^{16} + a^4\, b^{20} + b^{24})$$

$$+ \; (a^{20} + a^{16}\, b^4 + a^{12}\, b^8 + a^8\, b^{12} + a^4\, b^{16} + b^{20})\, c^4$$

$$+ \; (a^{16} + a^{12}\, b^4 + a^8\, b^8 + a^4\, b^{12} + b^{16})\, c^8$$

$$+ \; (a^{12} + a^8\, b^4 + a^4\, b^8 + b^{12})\, c^{12}$$

$$+ \; (a^8 + a^4\, b^4 + b^8)\, c^{16}$$

$$+ \; (a^4 + b^4)\, c^{20}$$

$$+ \quad c^{24} \Big\}.$$

$$B_7 = (a^{28} + a^{24}\, b^4 + a^{20}\, b^8 + a^{16}\, b^{12} + a^{12}\, b^{16} + a^8\, b^{20} + a^4\, b^{24} + b^{28})\, c^4$$

$$+ \; (a^{24} + a^{20}\, b^4 + a^{16}\, b^8 + a^{12}\, b^{12} + a^8\, b^{16} + a^4\, b^{20} + b^{24})\, c^8$$

$$+ \; (a^{20} + a^{16}\, b^4 + a^{12}\, b^8 + a^8\, b^{12} + a^4\, b^{16} + b^{20})\, c^{12}$$

$$+ \; (a^{16} + a^{12}\, b^4 + a^8\, b^8 + a^4\, b^{12} + b^{16})\, c^{16}$$

$$+ \; (a^{12} + a^8\, b^4 + a^4\, b^8 + b^{12})\, c^{20}$$

$$+ \; (a^8 + a^4\, b^4 + b^8)\, c^{24}$$

$$+ \; (a^4 + b^4)\, c^{28}.$$

$$+ \; (a^{24} + a^{20}\, b^4 + a^{16}\, b^8 + a^{12}\, b^{12} + a^8\, b^{16} + a^4\, b^{20} + b^{24})\, a^4\, b^4$$

$$+ \; (a^{20} + a^{16}\, b^4 + a^{12}\, b^8 + a^8\, b^{12} + a^4\, b^{16} + b^{20})\, a^4\, b^4\, c^4$$

$$+ \; (a^{16} + a^{12}\, b^4 + a^8\, b^8 + a^4\, b^{12} + b^{16})\, a^4\, b^4\, c^8$$

$$+ \; (a^{12} + a^8\, b^4 + a^4\, b^8 + b^{12})\, a^4\, b^4\, c^{12}$$

$$+ \; (a^8 + a^4\, b^4 + b^8)\, a^4\, b^4\, c^{16}$$

$$+ \; (a^4 + b^4)\, a^4\, b^4\, c^{20}$$

$$+ \quad a^4\, b^4\, c^{24}.$$

$$C_7 = (a^{28} + a^{24} b^4 + a^{20} b^8 + a^{16} b^{12} + a^{12} b^{16} + a^8 b^{20} + a^4 b^{24} + b^{28})$$
$$+ (a^{24} + a^{20} b^4 + a^{16} b^8 + a^{12} b^{12} + a^8 b^{16} + a^4 b^{20} + b^{24}) c^4$$
$$+ (a^{20} + a^{16} b^4 + a^{12} b^8 + a^8 b^{12} + a^4 b^{16} + b^{20}) c^8$$
$$+ (a^{16} + a^{12} b^4 + a^8 b^8 + a^4 b^{12} + b^{16}) c^{12}$$
$$+ (a^{12} + a^8 b^4 + a^4 b^8 + b^{12}) c^{16}$$
$$+ (a^8 + a^4 b^4 + b^8) c^{20}$$
$$+ (a^4 + b^4) c^{24}$$
$$+ c^{28}$$

In A_7, there are $\dfrac{1}{2} m(m-1)$ terms $= 28$.

" B_7 " " (m^2-1) " $= 63$.

" C_7 " " $\dfrac{1}{2} m(m+1)$ " $= 36$.

The reduction to the sphere will, therefore, give the identical equation,

$$1 - 36 + 63 - 28 = 0.$$

(7.)—Let $m = 9$.

The equation between the first 20 radii and perpendiculars will be,

$$p^2_{\,9}\ p^2_{\,8}\ p^2_{\,7}\ p^2_{\,6}\ p^2_{\,5}\ p^2_{\,4}\ p^2_{\,3}\ p^2_{\,2}\ p_i^2\ p^2\ r^2_{\,9}\ r^2_{\,8}\ r^2_{\,7}\ r^2_{\,6}\ r^2_{\,5}\ r^2_{\,4}\ r^2_{\,3}\ r^2_{\,2}\ r_i^2\ r^2$$

$$- C_8\ p^2_{\,9}\ p^2_{\,8}\ r^2_{\,9}\ r^2_{\,8}\ +\ B_8\ p^2_{\,9}\ r^2_{\,9}\ -\ A_8\ =\ 0.$$

$$A_8 = a^4 b^4 c^4 \Big\{ (a^{28} + a^{24} b^4 + a^{20} b^8 + a^{16} b^{12} + a^{12} b^{16} + a^8 b^{20} + a^4 b^{24} + b^{28}) c^4$$
$$+ (a^{24} + a^{20} b^4 + a^{16} b^8 + a^{12} b^{12} + a^8 b^{16} + a^4 b^{20} + b^{24}) c^8$$
$$+ (a^{20} + a^{16} b^4 + a^{12} b^8 + a^8 b^{12} + a^4 b^{16} + b^{20}) c^{12}$$
$$+ (a^{16} + a^{12} b^4 + a^8 b^8 + a^4 b^{12} + b^{16}) c^{16}$$
$$+ (a^{12} + a^8 b^4 + a^4 b^8 + b^{12}) c^{20}$$
$$+ (a^8 + a^4 b^4 + b^8) c^{24}$$
$$+ (a^4 + b^4) c^{28}$$

$$
\begin{aligned}
B_8 =\ &(a^{32}+a^{28}b^4+a^{24}b^6+a^{20}b^{12}+a^{16}b^{16}+a^{12}b^{20}+a^8 b^{24}+a^4 b^{28}+b^{32})\,c^4 \\
&+ (a^{28}+a^{24}b^4+a^{20}b^8+a^{16}b^{12}+a^{12}b^{16}+a^6 b^{20}+a^4 b^{24}+b^{28})\,c^6 \\
&+ (a^{24}+a^{20}b^4+a^{16}b^8+a^{12}b^{12}+a^8 b^{16}+a^4 b^{20}+b^{24})\,c^{12} \\
&+ (a^{20}+a^{16}b^4+a^{12}b^8+a^8 b^{12}+a^4 b^{16}+b^{20})\,c^{16} \\
&+ (a^{16}+a^{12}b^4+a^8 b^8+a^4 b^{12}+b^{16})\,c^{20} \\
&+ (a^{12}+a^8 b^4+a^4 b^8+b^{12})\,c^{24} \\
&+ (a^8+a^4 b^4+b^8)\,c^{28} \\
&+ (a^4+b^4)\,c^{32} \\
&+ (a^{28}+a^{24}b^4+a^{20}b^8+a^{16}b^{12}+a^{12}b^{16}+a^8 b^{20}+a^4 b^{24}+b^{28})\,a^4 b^4 \\
&+ (a^{24}+a^{20}b^4+a^{16}b^8+a^{12}b^{12}+a^8 b^{16}+a^4 b^{20}+b^{24})\,a^4 b^4 c^4 \\
&+ (a^{20}+a^{16}b^4+a^{12}b^8+a^8 b^{12}+a^4 b^{16}+b^{20})\,a^4 b^4 c^8 \\
&+ (a^{16}+a^{12}b^4+a^8 b^8+a^4 b^{12}+b^{16})\,a^4 b^4 c^{12} \\
&+ (a^{12}+a^8 b^4+a^4 b^8+b^{12})\,a^4 c^4 c^{16} \\
&+ (a^8+a^4 b^4+b^8)\,a^4 b^4 c^{20} \\
&+ (a^4+b^4)\,a^4 b^4 c^{24} \\
&+ a^4 b^4 c^{28}
\end{aligned}
$$

$$
\begin{aligned}
C_8 =\ &(a^{32}+a^{28}b^4+a^{24}b^8+a^{20}b^{12}+a^{16}b^{16}+a^{12}b^{20}+a^8 b^{24}+a^4 b^{28}+b^{32}) \\
&+ (a^{28}+a^{24}b^4+a^{20}b^8+a^{16}b^{12}+a^{12}b^{16}+a^8 b^{20}+a^4 b^{24}+b^{28})\,c^4 \\
&+ (a^{24}+a^{20}b^4+a^{16}b^8+a^{12}b^{12}+a^8 b^{16}+a^4 b^{20}+b^{24})\,c^8 \\
&+ (a^{20}+a^{16}b^4+a^{12}b^8+a^8 b^{12}+a^4 b^{16}+b^{20})\,c^{12} \\
&+ (a^{16}+a^{12}b^4+a^8 b^8+a^4 b^{12}+b^{16})\,c^{16} \\
&+ (a^{12}+a^8 b^4+a^4 b^8+b^{12})\,c^{20} \\
&+ (a^8+a^4 b^4+b^8)\,c^{24} \\
&+ (a^4+b^4)\,c^{28} \\
&+ c^{32}
\end{aligned}
$$

(XII.) In concluding this part of the subject, it is to be remarked that the co-efficients A_{m-1}, B_{m-1}, C_{m-1}, which have been employed in these expansions, are subject to three separate conditions; which may be incorporated into a single equation.

On examination of (1), (XI), page 82, we observe the relation,

$$c^4 B_{,} = (C_{,} - c^4) c^8 + A_{,}.$$

$$\therefore \quad A_{,} - c^4 B_{,} + c^8 C_{,} = c^{12}.$$

The axes being symmetrically combined in all the equations, it follows that,

$$A_{,} - b^4 B_{,} + b^8 C_{,} = b^{12}.$$

$$A_{,} - a^4 B_{,} + a^8 C_{,} = a^{12}.$$

consequently, since $\Sigma(a^0) = 3$,

$$A_{,} \Sigma(a^0) - B_{,} \Sigma(a^4) + C_{,} \Sigma(a^8) = \Sigma(a^{12}).$$

In a similar way will be found the relations,

$$A_2 - c^4 B_2 + c^8 C_2 = c^{16}.$$

$$A_2 - b^4 B_2 + b^8 C_2 = b^{16}.$$

$$A_2 - a^4 B_2 + a^8 C_2 = a^{16}.$$

$$A_2 \Sigma(a^0) - B_2 \Sigma(a^4) + C_2 \Sigma(a^8) = \Sigma(a^{16}).$$

In the general case there are the relations,

$$A_{m-1} - a^4 B_{m-1} + a^8 C_{m-1} = a^{4(m+1)}.$$

$$A_{m-1} - b^4 B_{m-1} + b^8 C_{m-1} = b^{4(m+1)}.$$

$$A_{m-1} - c^4 B_{m-1} + c^8 C_{m-1} = c^{4(m+1)}.$$

$$A_{m-1} \Sigma(a^0) - B_{m-1} \Sigma(a^4) + C_{m-1} \Sigma(a^8) = \Sigma\left\{a^{4(m+1)}\right\}. \quad (1)$$

From the last equation may be derived a verification of the preceding expansions. For, if the surface is reduced to a sphere, the general co-efficients, page 81, will severally take the values,

$$A_{m-1} = \frac{m(m-1)}{2} \frac{1}{a_{m+1}} = \frac{m(m-1)}{2} a^{4(m+1)}.$$

$$B_{m-1} = (m^2-1) \frac{1}{a_m} = (m^2-1) a^{4m}.$$

$$C_{m-1} = \frac{m(m+1)}{2} \frac{1}{a_{m-1}} = \frac{m(m+1)}{2} a^{4(m-1)}.$$

The introduction of these expressions into the equation (1) will give the identity,

$$\frac{m}{2}(m-1) - (m^2-1) + \frac{m}{2}(m+1) - 1 = 0.$$

From these equations of condition, which obtain among the co-efficients, it appears that in each series any one of them may be eliminated in terms of the remaining co-efficients; but this elimination leads to no new combinations.

CHAPTER V.

(I.) The investigation has hitherto, under an assumed hypothesis, exhibited to a certain extent connections existing between lines which belong to a given surface. In the present chapter it is further proposed to examine some of the relations connecting a system of surfaces generated from the primitive according to a given law.

Let it be conceived that a second ellipsoid is drawn through the point P_i, fig. 2, so situated and of such proportions that the first vectorial radius r shall be perpendicular to the tangent plane in contact with the second surface at P_i; it is required to determine this second surface.

The directions of the three principal planes in each ellipsoid are assumed to be coincident.

Let a_i, b_i, c_i be the semi-axes of the second surface; then, since P_i is situated upon it, if x_i, y_i, z_i are co-ordinates of that point,

$$\frac{x_i^2}{a_i^2} \quad + \quad \frac{y_i^2}{b_i^2} \quad + \quad \frac{z_i^2}{c_i^2} \quad = \quad 1. \qquad (1)$$

Now, if $\xi = \mu\zeta$, $\eta = \nu\zeta$, represent a perpendicular drawn from the centre upon a tangent plane of any ellipsoid, given by the equation,

$$\frac{x^2}{a^2} \quad + \quad \frac{y^2}{\beta^2} \quad + \quad \frac{z^2}{\gamma^2} \quad = \quad 1;$$

we have,

$$\mu = \frac{\gamma^2 x}{a^2 z}, \qquad \nu = \frac{\gamma^2 y}{\beta^2 z}.$$

Since the line r passes through the point $x\,y\,z$ upon the given surface, it will be represented by the equations,

$$\xi \;=\; \frac{x}{z}\zeta\,; \qquad \eta \;=\; \frac{y}{z}\zeta\,;$$

consequently, if r is at right angles to the plane which touches the second surface at the point $x_i\, y_i\, z_i$ we shall have,

$$\frac{x}{z} \;=\; \frac{c_i^2 x_i}{a_i^2 z_i}. \qquad \frac{y}{z} \;=\; \frac{c_i^2 y_i}{b_i^2 z_i}. \tag{2}$$

Further, since $x_i\, y_i\, z_i$ are co-ordinates of P_i, which is a point in the perpendicular upon a tangent plane at $x\,y\,z$ in the first surface, there will ensue these equations of relation,

$$\frac{x_i}{z_i} \;=\; \frac{c^2 x}{a^2 z}. \qquad \frac{y_i}{z_i} \;=\; \frac{c^2 y}{b^2 z}. \tag{3}$$

By the combination of (2) and (3) there is,

$$\frac{x}{z} \;=\; \frac{c_i^2}{a_i^2} \cdot \frac{c^2 x}{a^2 z}\,; \qquad \frac{y}{z} \;=\; \frac{c_i^2}{b_i^2} \cdot \frac{c^2 y}{b^2 z}\,;$$

consequently,

$$a\,a_i \;=\; b\,b_i \;=\; c\,c_i. \tag{4}$$

These are the relations between the axes of the two ellipsoids, shewing that they are reciprocally proportional.

(II.)　In the next place we have to ascertain the absolute values of the semi-axes $a_i\, b_i\, c_i$; it is evident that these lines will be functions of $a\,b\,c$, and of the co-ordinates $x\,y\,z$.

From page 2, there is,

$$x_i \;=\; \left(\frac{u}{a}\right)^2 x\,; \qquad y_i \;=\; \left(\frac{u}{b}\right)^2 y\,; \qquad z_i \;=\; \left(\frac{u}{c}\right)^2 z\,;$$

$$\frac{1}{u^4} \;=\; \Sigma\left(\frac{x^2}{a^6}\right);$$

And,
$$\frac{x_{i}^{2}}{a_{i}^{2}} + \frac{y_{i}^{2}}{b_{i}^{2}} + \frac{z_{i}^{2}}{c_{i}^{2}} = 1;$$

$$\left(\frac{u}{a}\right)^{4}\frac{x^{2}}{a_{i}^{2}} + \left(\frac{u}{b}\right)^{4}\frac{y^{2}}{b_{i}^{2}} + \left(\frac{u}{c}\right)^{4}\frac{z^{2}}{c_{i}^{2}} = 1;$$

consequently, eliminating by (4), (I),

$$\frac{1}{a^{2}a_{i}^{2}}\left\{\frac{x^{2}}{a^{2}} + \frac{y^{2}}{b^{2}} + \frac{z^{2}}{c^{2}}\right\} = \frac{1}{u^{4}}:$$

$$\therefore \quad u^{2} = aa_{i} = bb_{i} = cc_{i}. \qquad (5)$$

From the foregoing equations we obtain,

$$a_{i} = \frac{u^{2}}{a}. \qquad b_{i} = \frac{u^{2}}{b}. \qquad c_{i} = \frac{u^{2}}{c}. \qquad \cdot \ (6)$$

It has been shewn, in page 2, that $u^{2} = pr_{i}$; so that the connection between the axes of the co-related ellipsoids, is given also by the equations,

$$a_{i} = \frac{pr_{i}}{a}; \qquad b_{i} = \frac{pr_{i}}{b}; \qquad c_{i} = \frac{pr_{i}}{c}; \qquad (7)$$

expressions determining in magnitude the axes of the derived surface which passes through the point P_{i}, and has the plane in contact with that point at right angles to r, the vectorial radius of the primitive.

COR. 1. From the equation (5) of this section it appears that the line u is a mean proportional to each symmetrically expressed pair of the six semi-axes; while, from (6) it is evident that the greatest and least axes in the two surfaces interchange their directions; the *least* axis of the derived surface coinciding in direction with the *greatest* axis of the primitive, and conversely. The mean axis retains the same direction in both.

COR. 2. From (7) the following proportions are obtained;

$$\frac{a_{i}}{r_{i}} = \frac{p}{a}. \qquad \frac{b_{i}}{r_{i}} = \frac{p}{b}. \qquad \frac{c_{i}}{r_{i}} = \frac{p}{c}.$$

o

(III.)   Again, let an ellipsoid be drawn through the point Q, fig. (2), in which the tangent-plane of the first ellipsoid, in contact with it at P, is met by a perpendicular from the centre.   Let the tangent-plane in contact with the new surface at Q cut OP, or r, at right angles ; then, if $_{,}x$ $_{,}y$ $_{,}z$ are the co-ordinates of the point Q, there will be obtained by writing $n = 2$ in the general co-ordinates of equation (6), page 15,

$$_{,}x = \frac{p^2}{a^2}x. \qquad _{,}y = \frac{p^3}{b^2}y. \qquad _{,}z = \frac{p^2}{c^2}z.$$

$$\therefore \quad \frac{_{,}x}{_{,}z} = \frac{c^2 x}{a^2 z} ; \qquad \frac{_{,}y}{_{,}z} = \frac{c^2 y}{b^2 z}.$$

Let $_{,}a$ $_{,}b$ $_{,}c$ be the semi-axes of the ellipsoid, generated under this hypothesis; then, as in the preceding case,

$$\frac{x}{z} = \frac{c^2}{_{,}a^2}\frac{_{,}x}{_{,}z} ; \qquad \frac{y}{z} = \frac{c^2}{_{,}b^2}\frac{_{,}y}{_{,}z} ;$$

$$\frac{x}{z} = \frac{c^2}{_{,}a^2}\frac{c^2 x}{a^2 z} ; \qquad \frac{y}{z} = \frac{c^2}{_{,}b^2}\frac{c^2 y}{b^2 z} ;$$

consequently,

$$a\,_{,}a = b\,_{,}b = c\,_{,}c.$$

(IV.)   Let it now be supposed that ellipsoids are consecutively derived in a series from one another, in the same manner as the surface drawn through P, was generated from the primitive; according to this arrangement, the tangent-plane of the second derivative will meet p at right angles, that of the third r, and so on : so that r will be perpendicular upon the tangent planes, of the 1st, 3rd, 5th ... derived surfaces, while p is perpendicular not only to the tangent plane of the original ellipsoid, but to those also of the 2nd, 4th, 6th ... derivatives.   Each surface is, therefore, considered as a primary with respect to that which is next to it in sequence; or, rather, to *all* its subordinate surfaces.

Let a_m b_m c_m be the semi-axes of the mth derived surface,

x_m y_m z_m the co-ordinates of contact with its tangent-plane.

From what precedes, the following primary relations will subsist between the axes of consecutive surfaces.

$$a\,a_{,} \;=\; b\,b_{,} \;=\; c\,c_{,}.$$

$$a_{,}\,a_{2} \;=\; b_{,}\,b_{2} \;=\; c_{,}\,c_{2}.$$

$$a_{2}\,a_{3} \;=\; b_{2}\,b_{3} \;=\; c_{2}\,c_{3}.$$

$$\cdots \qquad \cdots$$

$$a_{m-1}\,a_{m} \;=\; b_{m-1}\,b_{m} \;=\; c_{m-1}\,c_{m}.$$

consequently,

$$a\,a_{m} \;=\; b\,b_{m} \;=\; c\,c_{m}. \tag{1}$$

For the co-ordinates of the several points of contact in the consecutive surfaces, taking into account only the positive sign, there will be,

$$x_{,} = \left(\frac{u}{a}\right)^{2} x. \qquad y_{,} = \left(\frac{u}{b}\right)^{2} y. \qquad z_{,} = \left(\frac{u}{c}\right)^{2} z.$$

$$x_{2} = \left(\frac{u_{,}}{a_{,}}\right)^{2} x_{,}. \qquad y_{2} = \left(\frac{u_{,}}{b_{,}}\right)^{2} y_{,}. \qquad z_{2} = \left(\frac{u_{,}}{c_{,}}\right)^{2} z_{,}.$$

$$x_{3} = \left(\frac{u_{2}}{a_{2}}\right)^{2} x_{2}. \qquad y_{3} = \left(\frac{u_{2}}{b_{2}}\right)^{2} y_{2}. \qquad z_{3} = \left(\frac{u_{2}}{c_{2}}\right)^{2} z_{2}.$$

$$\cdots \qquad \cdots \qquad \cdots \qquad \cdots$$

$$x_{m} = \left(\frac{u_{m-1}}{a_{m-1}}\right)^{2} x_{m-1}. \quad y_{m} = \left(\frac{u_{m-1}}{b_{m-1}}\right)^{2} y_{m-1}. \quad z_{m} = \left(\frac{u_{m-1}}{c_{m-1}}\right)^{2} z_{m-1}.$$

In these expressions it is necessary to be observed that the symbols $x_{m}\,y_{m}\,z_{m}$ and u_{m} are used in a sense differing from, although analogous to, that which has been assigned to them in the developments which have preceded. Hitherto they have been applied in reference to *consecutive* points on the *same* surface, but they now indicate the position of similar points on *consecutive* surfaces. There will then be the relation,

$$\frac{1}{u^{4}_{m}} \;=\; \frac{x^{2}_{m}}{a^{6}_{m}} \;+\; \frac{y^{2}_{m}}{b^{6}_{m}} \;+\; \frac{z^{2}_{m}}{c^{6}_{m}}. \tag{2}$$

Now, $u, u_2 \ldots u_m$ are the circle-ordinates in consecutive surfaces, corresponding in their character with u, in the original surface. We have then, as stated above,

$$x_m = \left(\frac{u_{m-1}}{a_{m-1}}\right)^2 x_{m-1}. \quad y_m = \left(\frac{u_{m-1}}{b_{m-1}}\right)^2 y_{m-1}. \quad z_m = \left(\frac{u_{m-1}}{b_{m-1}}\right)^2 z_{m-1}.$$

Consequently, the co-ordinates of the point of contact on the mth surface are given by the expressions,

$$x_m = \left\{\frac{u_{m-1}}{a_{m-1}} \frac{u_{m-2}}{a_{m-2}} \cdots \frac{u}{a}\right\}^2 x.$$

$$y_m = \left\{\frac{u_{m-1}}{b_{m-1}} \frac{u_{m-2}}{b_{m-2}} \cdots \frac{u}{b}\right\}^2 y.$$

$$z_m = \left\{\frac{u_{m-1}}{c_{m-1}} \frac{u_{m-2}}{c_{m-2}} \cdots \frac{u}{c}\right\}^2 z.$$

In these expressions it should be noticed that m may assume all integral values, from unity; and that u_0, a_0, &c., indicate u and a, as in former cases.

Each of the terms in the fractions included in x_m y_m z_m contains m factors.

For the reduction of these co-ordinates, we find, page 97,

$$a_1 = \frac{u^2}{a}; \quad b_1 = \frac{u^2}{b}; \quad c_1 = \frac{u^2}{c};$$

$$a_2 = \frac{u_1^2}{a_1}; \quad b_2 = \frac{u_1^2}{b_1}; \quad c_2 = \frac{u_1^2}{c_1};$$

$$\cdots \qquad \cdots \qquad \cdots \qquad \cdots \qquad \cdots$$

$$a_m = \frac{u^2_{m-1}}{a_{m-1}}; \quad b_m = \frac{u^2_{m-1}}{b_{m-1}}; \quad c_m = \frac{u^2_{m-1}}{c_{m-1}};$$

and, by consequence, if the several connecting equations are combined,

$$\frac{u^2{}_{m}-1 \;\; u^2{}_{m}-2 \;\; \cdots \;\; u^2{}_{,} \; u^2}{a_{m}-1 \;\; a_{m}-2 \;\; \cdots \;\; a_{,} \; a} \;=\; a_{m} \; a_{m}-1 \;\; \cdots \;\; a_2 \; a_{,}$$

$$\therefore \;\; \left\{ \frac{u_{m}-1 \;\; u_{m}-2 \;\; \cdots \;\; u_{,} \; u}{a_{m}-1 \;\; a_{m}-2 \;\; \cdots \;\; a_{,} \; a} \right\}^2 \;=\; \frac{a_{m}}{a}.$$

$$\left. \begin{array}{ccc} \therefore & x_{m} & = & \dfrac{a_{m}}{a} \; x. \\[2mm] & y_{m} & = & \dfrac{b_{m}}{b} \; y. \\[2mm] & z_{m} & = & \dfrac{c_{m}}{c} \; z. \end{array} \right\} \qquad (3)$$

the same form of expression being necessarily applicable to each of the axes.

From (3), $\quad \Sigma \left\{ \dfrac{x_{m}}{a_{m}} \right\}^2 \;=\; \Sigma \left\{ \dfrac{x}{a} \right\}^2 ;$

which is obviously true.

(V.) In following out this investigation, the next consideration that presents itself is the necessity of ascertaining the value of the lines $x_{m} \; y_{m} \; z_{m}$, expressed in terms of the original co-ordinates; but, before entering upon this inquiry, it may be worth while to notice that a singular relation obtains between the consecutive circle-ordinates $u, u_{,}, u_2, \ldots u_{m}$, and the co-ordinates of the generating point, $x \; y \; z$; combined with the axes of the given surface. From what has preceded, there will ensue the series of equations,

$$\frac{1}{u^4} \;=\; \left(\frac{x}{a}\right)^2 \frac{1}{a^4} \;+\; \left(\frac{y}{b}\right)^2 \frac{1}{b^4} \;+\; \left(\frac{z}{c}\right)^2 \frac{1}{c^4}.$$

$$\frac{1}{u_{,}^4} \;=\; \left(\frac{x}{a}\right)^2 \frac{1}{a_{,}^4} \;+\; \left(\frac{y}{b^2}\right) \frac{1}{b_{,}^4} \;+\; \left(\frac{z}{c}\right)^2 \frac{1}{c_{,}^4}.$$

$$\cdots \qquad \cdots \qquad \cdots \qquad \cdots$$

$$\frac{1}{u^4{}_{m}} \;=\; \left(\frac{x}{a}\right)^2 \frac{1}{a^4{}_{m}} \;+\; \left(\frac{y}{b}\right)^2 \frac{1}{b^4{}_{m}} \;+\; \left(\frac{z}{c}\right)^2 \frac{1}{c^4{}_{m}}.$$

RELATED ELLIPSOIDS.

Hence the equation is derived,

$$\Sigma\left\{\frac{1}{u^4}\right\} = \frac{x^2}{a^2}\,\Sigma\left\{\frac{1}{a^4}\right\} + \frac{y^2}{b^2}\,\Sigma\left\{\frac{1}{b^4}\right\} + \frac{z^2}{c^2}\,\Sigma\left\{\frac{1}{c^4}\right\};$$

in which it must be understood that all the lines are referred to a numerical unit.

The symbol Σ is here used in a sense different from that which was assigned to it in page 7, being now taken to indicate the sum of an unlimited series of analogous quantities.

(VI.) It now becomes necessary to ascertain the precise value of the axes in these consecutive surfaces; and, in order to this, we have to notice the equivalents, page 100,

$$\frac{1}{a_{,}} = \frac{a}{u^2}; \quad \frac{1}{b_{,}} = \frac{b}{u^2}; \quad \frac{1}{c_{,}} = \frac{c}{u^2};$$

$$\frac{1}{a_2} = \frac{a_{,}}{u_{,}^2}; \quad \frac{1}{b_2} = \frac{b_{,}}{u_{,}^2}; \quad \frac{1}{c_2} = \frac{c_{,}}{u_{,}^2};$$

$$\cdots$$

$$\frac{1}{a_{\mathrm{m}}} = \frac{a_{\mathrm{m}-1}}{u^2{}_{\mathrm{m}-1}}; \quad \frac{1}{b_{\mathrm{m}}} = \frac{b_{\mathrm{m}-1}}{u^2{}_{\mathrm{m}-1}}; \quad \frac{1}{c_{\mathrm{m}}} = \frac{c_{\mathrm{m}-1}}{u^2{}_{\mathrm{m}-1}};$$

From page 99,

$$\frac{1}{u_{,}^4} = \Sigma\left\{\frac{x_{,}^2}{a_{,}^6}\right\} = \frac{a^2\,x^2 + b^2\,y^2 + c^2\,z^2}{u^6}.$$

Assume v to be a line such that,

$$v^4 = a^2\,x^2 + b^-\,y^2 + c^2\,z^2,$$

then there is the relation,

$$u_{,} = \frac{u^2}{v}.$$

It appears, therefore, that u is a mean proportional to $u_{,}$ and v.

But u is also a mean proportional to p and $r_{,}$, page 2, (6); so that there is the geometrical relation,

$$u_{,} v \;=\; p\,r_{,}$$

or, $$u_{,} \;:\; r_{,} \;::\; p \;:\; r.$$

The generalising of this investigation will lead to the following series of values for the circle-ordinates in the consecutive ellipsoids, by which they are expressed in terms of the rectangular co-ordinates of the point assumed upon the original surface.

$$\frac{1}{u^{4}} \;=\; \Sigma \left\{ \frac{x^{2}}{a^{6}} \right\}.$$

$$u_{,} \;=\; \frac{u^{2}}{v}.$$

$$u_{2} \;=\; \frac{u^{3}}{v^{2}}.$$

$$u_{3} \;=\; \frac{u^{4}}{v^{3}}.$$

$$\cdots$$

$$u_{m} \;=\; \frac{u^{m+1}}{v^{m}}.$$

COR. The ratio of the circle-ordinates in the consecutive ellipsoids is constant, for we have,

$$\frac{u}{u_{,}} \;=\; \frac{u_{,}}{u_{2}} \;=\; \frac{u_{2}}{u_{3}} \;=\; \cdots \;=\; \frac{u_{m}}{u_{m+1}} \;=\; \frac{v}{u}.$$

(VII.) In the determination of the axes of the several surfaces we find,

$$a_{0} = a; \qquad b_{0} = b; \qquad c_{0} = c;$$

$$a_{2} = \left(\frac{u}{v}\right)^{2} a; \quad b^{2} = \left(\frac{u}{v}\right)^{2} b; \quad c_{2} = \left(\frac{u}{b}\right)^{c} c;$$

$$a_{4} = \left(\frac{u}{v}\right)^{4} a; \quad b_{4} = \left(\frac{u}{v}\right)^{4} b; \quad c_{4} = \left(\frac{u}{v}\right)^{4} c;$$

$$\cdots \qquad \cdots \qquad \cdots \qquad \cdots \qquad \cdots$$

$$a_{2m} = \left(\frac{u}{v}\right)^{2m} a; \quad b_{2m} = \left(\frac{u}{v}\right)^{2m} a; \quad c_{2m} = \left(\frac{u}{v}\right)^{2m} c;$$

$$a_i = \frac{u^2}{v^0}\frac{1}{a}. \qquad b_i = \frac{u^2}{v^0}\frac{1}{b}. \qquad c_i = \frac{u^2}{v^0}\frac{1}{c}.$$

$$a_3 = \frac{u^4}{v^2}\frac{1}{a}. \qquad b_3 = \frac{u^4}{v^2}\frac{1}{b}. \qquad c_3 = \frac{u^4}{v^2}\frac{1}{c}.$$

$$a_5 = \frac{u^6}{v^4}\frac{1}{a}. \qquad b_5 = \frac{u^6}{v^4}\frac{1}{b}. \qquad c_5 = \frac{u^6}{v^4}\frac{1}{c}.$$

$$\cdots \qquad\qquad \cdots \qquad\qquad \cdots \qquad\qquad \cdots$$

$$a_{2m-1} = \frac{u^{2m}}{v^{2(m-1)}}\frac{1}{a}. \qquad b_{2m-1} = \frac{u^{2m}}{v^{2(m-1)}}\frac{1}{b}. \qquad c_{2m-1} = \frac{u^{2m}}{v^{2(m-1)}}\frac{1}{c}.$$

In the expressions for the axes of the alternate surfaces, commencing with the given ellipsoid, any integral value may be assigned to m; and, in those which belong to the alternate surfaces beginning from the first derivative, all integral values may be taken for m, from one to infinity.

Let $\Sigma(a_0)$ and $\Sigma(a_i)$ respectively represent the sums of the semi-axes, in the two series of surfaces taken alternately, then, employing Σ to represent the sum of any number of these lines,

$$\Sigma(a_0) = a\left\{1 + \frac{u^2}{v^2} + \frac{u^4}{v^4} + \ldots\right\}.$$

$$\Sigma(a_i) = \frac{u^2}{a}\left\{1 + \frac{u^2}{v^2} + \frac{u^4}{v^4} + \ldots\right\}.$$

The summation of these geometrical series to n terms will give the results,

$$\Sigma(a_0) = \frac{a(u^{2n} - v^{2n})}{(u^2 - v^2)\,v^{2n-2}}.$$

$$\Sigma(a_i) = \frac{u^2(u^{2n} - v^{2n})}{a(u^2 - v^2)\,v^{2n-2}}.$$

consequently,

$$\frac{\Sigma(a_i)}{\Sigma(a_0)} = \left(\frac{u}{a}\right)^2.$$

Further, since $\dfrac{u}{v}$ is necessarily a fraction less than unity,* except at the extremities of each of the axes, at which points it becomes equal to unity, the two series are convergent, and their summation may be taken for an infinite number of terms; so that,

$$\Sigma\,(a_0) \;=\; \frac{a\,v^2}{v^2 - u^2}.$$

$$\Sigma\,(a_{\prime}) \;=\; \frac{u^2\,v^2}{a\,(v^2 - u^2)} \;=\; \frac{a_{\prime}\,v^2}{v^2 - u^2}.$$

Now, since, $\dfrac{u^2}{a} = a_{\prime}$, $\therefore \left(\dfrac{u}{a}\right)^3 = \dfrac{a_{\prime}}{a}$; consequently,

$$\frac{\Sigma\,(a_{\prime})}{\Sigma\,(a_0)} \;=\; \frac{a_{\prime}}{a}:$$

$$\frac{\Sigma\,(b_{\prime})}{\Sigma\,(b_0)} \;=\; \frac{b_{\prime}}{b}:$$

$$\frac{\Sigma\,(c_{\prime})}{\Sigma\,(c_0)} \;=\; \frac{c_{\prime}}{c}:$$

the same forms being necessarily true for each series of axial lines.

(VIII.) From the values assigned to the several systems of axes in (VII), combined with the equations (3), page 101, it results that:

---

* NOTE.—Assuming that $\dfrac{u}{v}$ is a proper fraction, we have,

$$\Sigma\,(a^2\,x^3)\;\; \Sigma\left(\frac{x^2}{a^6}\right) \;>\; 1;$$

and, consequently, after reduction,

$$a^6\,(b^4 - c^4)^2\,y^2\,z^2 \;+\; b^6\,(a^4 - c^4)^2\,x^2\,z^2 \;+\; c^6\,(a^4 - b^4)^2\,x^2\,y^2 \;>\; 0:$$

which it is clear will be generally true.

P

1st. If m is any *even* number, as 0, 2, 4,   ...

$$x_m = \left(\frac{u}{v}\right)^m x. \qquad y_m = \left(\frac{u}{v}\right)^m y. \qquad z_m = \left(\frac{u}{v}\right)^m z.$$

2nd. If m is any *uneven* number, as 1, 3, 5,   ...

$$x_m = \left(\frac{u}{v}\right)^{m+1}\left(\frac{v}{a}\right)^2 x. \qquad y_m = \left(\frac{u}{v}\right)^{m+1}\left(\frac{v}{b}\right)^2 y. \qquad z_m = \left(\frac{u}{v}\right)^{m+1}\left(\frac{v}{c}\right)^2 z.$$

(IX.)   Let the sum of the volumes of the two infinite series of ellipsoids, which have been here considered, each series having reference to alternate surfaces, be represented under the symbol Σ ; then

$$\Sigma(V_o) = \frac{4\pi}{3}\left\{ a\,b\,c + a_2\,b_2\,c_2 + \dots \right\}.$$

$$\Sigma(V_i) = \frac{4\pi}{3}\left\{ a_i\,b_i\,c_i + a_3\,b_3\,c_3 + \dots \right\}.$$

After these series have been represented in terms of the values of the consecutive axes, as given in pages 103 and 104, there will be found,

$$\Sigma(V_o) = \frac{4\pi abc}{3}\left\{ 1 + \frac{u^{12}}{v^{12}} + \dots \right\} = \frac{4\pi abc}{3}\,\frac{v^{12}}{v^{12}-u^{12}}.$$

$$\Sigma(V_i) = \frac{4\pi u^6}{3\,abc}\left\{ 1 + \frac{u^6}{v^6} + \dots \right\} = \frac{4\pi}{3\,abc}\,\frac{u^6\,v^6}{v^6-u^6}.$$

Hence may be derived, involving the value of π, the singular expression,

$$\frac{u^3}{v^3}\,(u^6+v^6)^{\frac{1}{3}}\pi = \frac{3}{4}\,V_o\,\frac{\Sigma^{\frac{1}{3}}(V_i)}{\Sigma^{\frac{1}{3}}(V_o)}.$$

or, $\pi = \dfrac{3}{4}\,\dfrac{V_o}{u^3}\,\dfrac{\Sigma^{\frac{1}{3}}(V_i)}{\Sigma^{\frac{1}{3}}(V_o)}\left\{ 1 - \dfrac{1}{2}\dfrac{u^6}{v^6} + \dfrac{3}{2^3}\dfrac{u^{12}}{v^{12}} - \dfrac{5}{2^4}\dfrac{u^{18}}{v^{18}} + \dots \right\}$

When V_0 is replaced by its proper value, the symbol π will disappear, leaving, as a relation between the two infinite series,

$$\frac{\Sigma^{\frac{1}{2}}(V_,)}{\Sigma^{\frac{1}{2}}(V_0)} = \frac{u^3}{v^3} \frac{(u^6 + v^6)^{\frac{1}{2}}}{a\,b\,c}.$$

(X.) In (I), page 95, figure 2, the first derivative has been assumed to pass through the point $P_,$, subject to the condition that its tangent plane at that point is intersected at right angles by the first vectorial radius. Let it be supposed that this method of derivation is generalised, so that ellipsoids are drawn to pass through each of the consecutive points $P_, P_2$ ... P_m under the hypothesis that the tangent plane of the surface passing through P_m is pierced at right angles by the radius r_{m-1}.

The seminaxes of the first derived surface being $a_, b_, c_,$ it has been already shewn, in page 97, that,

$$a_, = \frac{u^2}{a}; \qquad b_, = \frac{u^2}{b}; \qquad c_, = \frac{u^2}{c};$$

and, by adapting the investigation to the second surface, it will be found that,

$$a_2 = \frac{u_,^2}{a}; \qquad b_2 = \frac{u_,^2}{b}; \qquad c_2 = \frac{u_,^2}{c};$$

Similarly,

$$a_3 = \frac{u_2^2}{a}; \qquad b_3 = \frac{u_2^2}{b}; \qquad c_3 = \frac{u_2^2}{c};$$

(XI.) In the general case we have, for the equation of the mth derived surface,

$$\frac{x^2_m}{a^2_m} + \frac{y^2_m}{b^2_m} + \frac{z^2_m}{c^2_m} = 1.$$

But, from page 8 (corrected),

$$x_m = \left(\frac{v_m}{a}\right)^{2m} x; \quad \&\text{c.}$$

Hence we may find,

$$a\, a_m \;=\; b\, b_m \;=\; c\, c_m \,;$$

and, consequently,

$$a_m = \frac{u^2_{m-1}}{a}. \qquad b_m = \frac{u^2_{m-1}}{b}. \qquad c_m = \frac{u^2_{m-1}}{c}.$$

There is, then, the following series of relations,

$$aa_{,} \;=\; bb_{,} \;=\; cc_{,}.$$

$$aa_{2} \;=\; bb_{2} \;=\; cc_{2}.$$

$$\cdots \qquad \cdots$$

$$aa_m \;=\; bb_m \;=\; cc_m.$$

(XII.)   The equations of the preceding section imply the conditions,

$$\frac{a_{,}}{a_{2}} \;=\; \frac{b_{,}}{b_{2}} \;=\; \frac{c_{,}}{c_{2}}.$$

$$\cdots \qquad \cdots$$

$$\frac{a_{m-1}}{a_m} \;=\; \frac{b_{m-1}}{b_m} \;=\; \frac{c_{m-1}}{c_m}.$$

The surfaces, therefore, which are generated under this hypothesis, are all similar to each other.

Further, we have,

$$\frac{a_{,}}{b_{,}} \;=\; \frac{b}{a}\,; \qquad \frac{a_{,}}{c_{,}} \;=\; \frac{c}{a}\,; \qquad \frac{b_{,}}{c_{,}} \;=\; \frac{c}{b}\,;$$

shewing that the axes of the first derivative, and, consequently, of all those which follow, are reciprocally proportional to the axes of the primary.

(XIII.) From the preceding considerations it may be shewn that,

$$\frac{1}{a^3{}_m} = a^2 \frac{\Sigma\left(\frac{x^2}{a^{4m+2}}\right)}{\Sigma\left(\frac{x^2}{a^{4m-2}}\right)}.$$

$$\frac{1}{b^2{}_m} = b^2 \frac{\Sigma\left(\frac{x^2}{a^{4m+2}}\right)}{\Sigma\left(\frac{x^2}{a^{4m-2}}\right)}.$$

$$\frac{1}{c^3{}_m} = c^2 \frac{\Sigma\left(\frac{x^2}{a^{4m+2}}\right)}{\Sigma\left(\frac{x^2}{a^{4m-2}}\right)}.$$

By these expressions the values of the axes, in each of the consecutive surfaces, are fully determined, as functions of the initial point of contact assumed upon the primary ellipsoid.

(XIV.) Reverting to the expressions in page 107, we observe the relations,

$$aa_, = u^2 = pr_,.$$

$$aa_, = u_,^2 = p_, r_,.$$

$$\dots \qquad \dots$$

$$a\,a_m = u^2{}_{m-1} = p_{m-1}\,r_m.$$

Consequently,

$$a_, : r_, :: p : a.$$

$$a_, : r_, :: p_, : a.$$

$$\dots \qquad \dots \qquad \dots \qquad \dots$$

$$a_m : r_m :: p_{m-1} : a.$$

Since $aa_m = bb_m = cc_m$ the proportions which have here been given will apply equally to any of the three axes.

(XV.) The proportions which have been educed in (XIV) exhibit in a peculiar light the character of the lines of which we have been considering some of the properties. It could hardly have been anticipated, and must appear in some degree remarkable, that these lines should be thus intimately allied to a series of consecutive ellipsoids; in fact, not less closely, perhaps, than they are connected with the surface to which they seem originally to belong. If it is further considered that each point of intersection on the primary is common to it with the corresponding derivative, while on every new derivative this may be assumed as a starting point for similar combinations in relation to that ellipsoid, there is a probability that an examination of the surfaces under this aspect would bring out not a few singular results. We have not, however, space at present to enter upon such an inquiry.

(XVI.) The relations which have been examined in this chapter, partial, as applying only to the ellipsoid, should now be considered with reference to the general surface, symbolised by the equation,

$$\Sigma \left(\frac{x^n}{a^n} \right) \;=\; 1. \tag{1}$$

Let a surface of the same order be supposed to pass through the point $P_{\prime}$, fig. 2, in such a manner that its tangent plane drawn to that point may meet the first vectorial radius O P at right angles. This derived surface may be represented by,

$$\Sigma \left(\frac{x_{\prime}^{\,n}}{a_{\prime}^{\,n}} \right) \;=\; 1. \tag{2}$$

the point $x_{\prime}, y_{\prime}, z_{\prime}$, as well as, of course, a certain line of double curvature, being common to it with the primary.

The equations of the tangent planes are, respectively,

$$\frac{x^{n-1}}{a^n} \, \xi \;+\; \frac{y^{n-1}}{b^n} \, \eta \;+\; \frac{z^{n-1}}{c^n} \, \zeta \;=\; 1.$$

$$\frac{x_{\prime}^{\,n-1}}{a_{\prime}^{\,n}} \, \xi \;+\; \frac{y_{\prime}^{\,n-1}}{b_{\prime}^{\,n}} \, \eta \;+\; \frac{z_{\prime}^{\,n-1}}{c_{\prime}^{\,n}} \, \zeta \;=\; 1.$$

From the given hypothesis we have, therefore,

$$\frac{x}{z} = \frac{c_i^n x_i^{n-1}}{a_i^n z_i^{n-1}}. \qquad \frac{y}{z} = \frac{c_i^n y_i^{n-1}}{b_i^n z_i^{n-1}}.$$

$$\frac{x_i}{z_i} = \frac{c^n x^{n-1}}{a^n z^{n-1}}. \qquad \frac{y_i}{z_i} = \frac{c^n y^{n-1}}{b^n z^{n-1}}.$$

After these conditions have been combined, there will be found the relations,

$$\left. \begin{array}{rcl} \dfrac{c_i}{a_i} \dfrac{c^{n-1} x^{n-2}}{a^{n-1} z^{n-2}} & = & 1. \\[3mm] \dfrac{c_i}{b_i} \dfrac{c^{n-1} y^{n-2}}{b^{n-1} z^{n-2}} & = & 1. \end{array} \right\} \qquad (3)$$

Cor. If the general surface is reduced to the ellipsoid, $n = 2$, and, as in page 108, these equations take the form,

$$a a_i = b b_i = c c_i;$$

it does not, however, appear that this simple relation is true with respect to any surface of different order, which can be symbolized by the form (1).

(XVII.) It is requisite, in the next place, to ascertain the absolute values of the new axes.

From page 15,

$$x_i = \frac{u^2 x^{n-1}}{a^n}. \qquad y_i = \frac{u^2 y^{n-1}}{b^n}. \qquad z_i = \frac{u^2 z^{n-1}}{c^n}.$$

By introducing these equivalents in (2), together with the expressions for the semi-axes a_i, b_i, obtained from (3), we shall find, after the reductions requisite,

$$\frac{z^{n(n-2)}}{c^{n(n-1)}} \Sigma \left(\frac{x^n}{a^n} \right) = \frac{c_i^n}{u^{2n}}.$$

Hence,

$$c_{,} = \frac{u^2 z^{n-2}}{c^{n-1}}.$$

consequently,

$$b_{,} = \frac{u^2 y^{n-2}}{b^{n-1}}.$$

$$a_{,} = \frac{u^2 x^{n-2}}{a^{n-1}}.$$

If the foregoing examination is extended to the second derived surface, it will be found that,

$$a_2 = \frac{u_{,}^2 x_{,}^{n-2}}{a^{n-1}}; \qquad b_2 = \frac{u_{,}^2 y_{,}^{n-2}}{b^{n-1}}; \qquad c_2 = \frac{u_{,}^2 z_{,}^{n-2}}{c^{n-1}};$$

but this is not necessary, since it is evident, from the nature of the case, that we may at once adopt the generalisation,

$$a_m = \frac{u^2_{m-1}\, x^{n-2}_{m-1}}{a^{n-1}}.$$

$$b_m = \frac{u^2_{m-1}\, y^{n-2}_{m-1}}{b^{n-1}}.$$

$$c_m = \frac{u^2_{m-1}\, z^{n-2}_{m-1}}{c^{n-1}}.$$

(XVIII.) By writing $m - 1$ for m in the expressions deduced in page 20, the symbol v being corrected as in the errata, it will be found that,

$$x_{m-1}^{\,n-2} = \left(\frac{v^2_{m-1}}{u^n}\right)^{(n-1)^{m-1} - 1} x^{(n-1)^{m-1}(n-2)}.$$

$$\therefore \quad a_m = \frac{u^2_{m-1}\,(v_{m-1})^{2(n-1)^{m-1} - 1}}{a^{n(n-1)^{m-1} - 1}}\, x^{(n-1)^{m-1}(n-2)}.$$

with similar expressions for b_m, c_m given by the appropriate substitutions.

These values of the axes in the mth surface may be expressed more symmetrically under the following form.

$$a_m = a \left(\frac{u_{m-1}}{v_{m-1}}\right)^2 \left(\frac{v_{m-1}}{x}\right)^{2(n-1)} \left(\frac{x}{a}\right)^{n(n-1)^{\frac{m-1}{}}}.$$

$$b_m = b \left(\frac{u_{m-1}}{v_{m-1}}\right)^2 \left(\frac{v_{m-1}}{y}\right)^{2(n-1)} \left(\frac{y}{b}\right)^{n(n-1)^{\frac{m-1}{}}}.$$

$$c_m = c \left(\frac{u_{m-1}}{v_{m-1}}\right)^2 \left(\frac{v_{m-1}}{z}\right)^{2(n-1)} \left(\frac{z}{c}\right)^{n(n-1)^{\frac{m-1}{}}}.$$

In these formulæ we have to notice, from page 20 (corrected), that,

$$\left(v_{m-1}\right)^{\frac{(n-1)-1}{n-2}^{\frac{m-1}{}}} = u_{m-2}\, u_{m-3}\, \ldots\, u_{\prime}\, u.$$

(XIX.) Some singular relations involving the circle-ordinates $u, u_{\prime}, \ldots$ may be here appropriately noticed.

Since

$$a_m = u^2_{m-1} \frac{x^{n-2}_{m-1}}{a^{n-1}}.$$

$$\therefore \quad \frac{a_m}{a} x^2_{m-1} = u^2_{m-1} \frac{x^n_{m-1}}{a^n}.$$

It is evident that similar relations are true for the remaining semi-axes $b_m\ c_m$.

But, $x_{m-1}\ y_{m-1}\ z_{m-1}$ is a point upon the primary surface;

$$\therefore \quad \Sigma \left(\frac{x^n_{m-1}}{a^n}\right) = 1.$$

From this it will follow that,

$$u^2_{m-1} = \frac{a_m}{a} x^2_{m-1} + \frac{b_m}{b} y^2_{m-1} + \frac{c_m}{c} z^2_{m-1}.$$

$$u^2_{m-2} = \frac{a_{m-1}}{a} x^2_{m-2} + \frac{b_{m-1}}{b} y^2_{m-2} + \frac{c_{m-1}}{c} z^2_{m-2}.$$

$$\ldots \qquad \ldots \qquad\qquad\qquad \ldots$$

Q

$$u_{\prime}^{2} \;=\; \frac{a_2}{a} x_{\prime}^{2} \;+\; \frac{b_2}{b} y_{\prime}^{2} \;+\; \frac{c_2}{c} z_{\prime}^{2}.$$

$$u^{2} \;=\; \frac{a_{\prime}}{a} x^{2} \;+\; \frac{b_{\prime}}{b} y^{2} \;+\; \frac{c_{\prime}}{c} z^{2}.$$

Again, from page 16, (III),

$$x_m \;=\; \frac{u^2_{m-1} \, x^{n-1}_{m-1}}{a^n} ; \quad \text{with similar expressions for } y_m \; z_m.$$

$$\therefore \; x_{m-1} \, x_m \;=\; u^2_{m-1} \frac{x^n_{m-1}}{a^n} : \;\; \&c.$$

Consequently,

$$x_{m-1} \, x_m \;+\; y_{m-1} \, y_m \;+\; z_{m-1} \, z_m \;=\; u^2_{m-1} \, \Sigma \left(\frac{x^n_{m-1}}{a^n} \right).$$

Hence we have,

$$u^2_{m-1} \;=\; x_{m-1} \, x_m \;+\; y_{m-1} \, y_m \;+\; z_{m-1} \, z_m.$$

$$u^2_{m-2} \;=\; x_{m-2} \, x_{m-1} \;+\; y_{m-2} \, y_{m-1} \;+\; z_{m-1} \, z_m.$$

$$\ldots \qquad \ldots \qquad \ldots$$

$$u^2 \;=\; x \, x_{\prime} \;+\; y \, y_{\prime} \;+\; z \, z_{\prime}.$$

In verification of these expressions for the circle-ordinates, since $x \, y \, z$ and $x_{\prime} \, y_{\prime} \, z_{\prime}$ are points taken upon the lines $r \, r_{\prime}$, which intersect at O; let i be the inclination of these lines, then,

$$r \, r_{\prime} \cos. \, i \;=\; x \, x_{\prime} \;+\; y \, y_{\prime} \;+\; z \, z_{\prime} \;=\; u^2 \;=\; p \, r_{\prime}.$$

$$\therefore \;\; \cos. \, i \;=\; \frac{p}{r}.$$

(XX.) In order to complete this part of the subject, the investigation commencing at (IV), page 98, as applied to the ellipsoid, should now be extended so as to include the general surface of the nth degree. In the figure (3) P P$_{\prime}$; P$_{\prime}$ P$_{2}$; ... are portions of surfaces drawn after the same law as the ellipsoids in the section to which reference has been made ; *i.e.* the primary surface P P$_{\prime}$ has O Q$_{\prime}$, or p, for the perpendicular upon its plane of contact at any given point P : again, O Q$_{\prime\prime}$, or r, is perpendicular to the plane which touches the first derivative P$_{\prime}$P$_{2}$ at the point P$_{\prime}$. Regarding the consecutive derivation of these surfaces from each other, according to this hypothesis, as unlimited, we have to ascertain their mutual relations. This will involve the determination of their axes, and of the co-ordinates to those points in which the consecutive surfaces are pierced by the lines p and r.

Now, the point P$_{2}$ in the surface P$_{\prime}$ P$_{2}$ is derived from P$_{\prime}$ in precisely the same way in which P$_{\prime}$ is derived from P in the given surfaces P P$_{\prime}$; and the same law of derivation applies to each of the following surfaces. Thus, if $u_{\prime}$ u_{2} u_{3} ... are lines in the consecutive surfaces of a character analogous to that of u in the primary, the co-ordinates of each successive point of intersection being expressed according to this principle, there will ensue the relations,

$$x_{\prime} \quad = \quad \frac{u^2}{a^n} \, x^{n-1}.$$

$$x_{2} \quad = \quad \frac{u_{\prime}^{2}}{a_{\prime}^{n}} \, x_{\prime}^{n-1}.$$

$$x_{3} \quad = \quad \frac{u^2_{2}}{a^n_{2}} \, x_{2}^{n-1}.$$

$$\ldots \qquad \ldots$$

$$x_m \quad = \quad \frac{u^2_{m-1}}{a^n_{m-1}} \, x^{n-1}_{m-1}.$$

By the reduction of these fundamental expressions there will be, in the general case,

$$x_m = \frac{\begin{array}{cccccc} 2 & 2(n-1) & 2(n-1)^2 & & 2(n-1)^{m-1} & \\ u_{m-1} & u_{m-2} & u_{m-3} & \ldots & u & (n-1)^m \\ \hline n & n(n-1) & n(u-1)^2 & & n(n-1)^{m-1} & \\ a_{m-1} & a_{m-2} & a_{m-3} & \ldots & a & \end{array}} x. \tag{1}$$

The co-ordinates $y_m\, z_m$ are included under the same form by the interchange of a and x with b and y, or c and z.

(XXI.) The consideration which is now presented to us consists in the determination of the absolute values of $x_m\, y_m\, z_m$, and of the axes $a_m\, b_m\, c_m$ which belong to the consecutive surfaces.

In page 111, it has been already shewn that, in the first derivative P, P_2,

$$\frac{c_i}{a_i}\frac{c^{n-1} x^{n-2}}{a^{n-1} z^{n-2}} = 1.$$
$$\frac{c_i}{b_i}\frac{c^{n-1} y^{n-2}}{b^{n-1} z^{n-2}} = 1. \Bigg\}$$

Again, the second derivative $P_2\, P_3$ is related to the first, P, P_2, in the same connection which this latter surface bears to the primary. We have then, representing the tangent plane of the second derived surface at P_2,

$$\frac{x_2^{n-1}}{a_2^{n}} \xi + \ldots = 1.$$

The equations of the perpendicular $O\,Q$ will be,

$$\xi = \frac{x_i}{z_i} \zeta. \qquad \eta = \frac{y_i}{z_i} \zeta.$$

consequently,

$$\frac{x_i}{z_i} = \frac{c^n_2\, x_2^{n-1}}{a^n_2\, z_2^{n-1}}. \qquad \frac{y_i}{z_i} = \frac{c^n_2\, y_2^{n-1}}{b^n_2\, z_2^{n-1}}.$$

This relation would necessarily arise from the law of mutual interdependence which has been assumed to unite the consecutive surfaces; and, by generalizing the application of this law, there will be found the following series of equalities.

On the line O Q,

$$\frac{x_{_{/}}}{z_{_{/}}} = \frac{c^n x^{n-1}}{a^n z^{n-1}}. \qquad\qquad \frac{y_{_{/}}}{z_{_{/}}} = \frac{c^n y^{n-1}}{b^n z^{n-1}}.$$

$$\frac{x_{_{/}}}{z_{_{/}}} = \frac{c_2{}^n x_2{}^{n-1}}{a_2{}^n z_2{}^{n-1}}. \qquad\qquad \frac{y_{_{/}}}{z_{_{/}}} = \frac{c_2{}^n y_2{}^{n-1}}{b_2{}^n z_2{}^{n-1}}.$$

$$\frac{x_{_{/}}}{z_{_{/}}} = \frac{c_4{}^n x_4{}^{n-1}}{a_4{}^n z_4{}^{n-1}}. \qquad\qquad \frac{y_{_{/}}}{z_{_{/}}} = \frac{c_4{}^n y_4{}^{n-1}}{b_4{}^n z_4{}^{n-1}}.$$

$$\dots \qquad \dots \qquad \dots \qquad \dots$$

Again, on the line OP,

$$\frac{x}{z} = \frac{c_{_{/}}{}^n x_{_{/}}{}^{n-1}}{a_{_{/}}{}^n z_{_{/}}{}^{n-1}}. \qquad\qquad \frac{y}{z} = \frac{c_{_{/}}{}^n y_{_{/}}{}^{n-1}}{b_{_{/}}{}^n z_{_{/}}{}^{n-1}}.$$

$$\frac{x}{z} = \frac{c_3{}^n x_3{}^{n-1}}{a_3{}^n z_3{}^{n-1}}. \qquad\qquad \frac{y}{z} = \frac{c_3{}^n y_3{}^{n-1}}{b_3{}^n z_3{}^{n-1}}.$$

$$\frac{x}{z} = \frac{c_5{}^n x_5{}^{n-1}}{a_5{}^n z_5{}^{n-1}}. \qquad\qquad \frac{y}{z} = \frac{c_5{}^n y_5{}^{n-1}}{b_5{}^n z_5{}^{n-1}}.$$

$$\dots \qquad \dots \qquad \dots \qquad \dots$$

The tangent planes of the alternate surfaces are, evidently, all parallel.

(XXII.) We have, in the next place, to consider the relations existing between the consecutive points of intersection.

Since the line O P passes through each of the points $(x\,y\,z)$, $(x_2\,y_2\,z_2)$, $(x_4\,y_4\,z_4)$, ... there will be the equations,

$$\frac{x}{z} = \frac{x_2}{z_2} = \frac{x_4}{z_4} = \dots$$

$$\frac{y}{z} = \frac{y_2}{z_2} = \frac{y_4}{z_4} = \dots$$

118 RELATED SURFACES.

Again, since the line O Q passes through the points (x, y, z_i), $(x_3\ y_3\ z_3)$, $(x_5\ y_5\ z_5)$ ... it is plain that,

$$\frac{x_i}{z_i} = \frac{x_3}{z_3} = \frac{x_5}{z_5} = \cdots$$

$$\frac{y_i}{z_i} = \frac{y_3}{z_3} = \frac{y_5}{z_5} = \cdots$$

(XXIII.) From the expressions given in (XXI) we have,

$$\frac{x_i}{z_i} = \left(\frac{c_2}{a_2}\right)^n \left(\frac{x_2}{z_2}\right)^{n-1} = \left(\frac{c_2}{a_2}\right)^n \left(\frac{x}{z}\right)^{n-1}$$

consequently,

$$\left(\frac{c}{a}\right)^n \left(\frac{x}{z}\right)^{n-1} = \left(\frac{c_2}{a_2}\right)^n \left(\frac{x}{z}\right)^{n-1}.$$

$$\therefore \frac{c_2}{a_2} = \frac{c}{a}.$$

similarly,

$$\frac{c_2}{b_2} = \frac{c}{b}.$$

Again, we have,

$$\frac{x}{z} = \left(\frac{c_3}{a_3}\right)^n \left(\frac{x_3}{z_3}\right)^{n-1} = \left(\frac{c_3}{a_3}\right)^n \left(\frac{x_i}{z_i}\right)^{n-1} = \left(\frac{c_3}{a_3}\right)^n \left(\frac{c}{a}\right)^{n(n-1)} \left(\frac{x}{z}\right)^{(n-1)^2}.$$

$$\therefore 1 = \frac{c_3}{a_3} \left(\frac{c}{a}\right)^{n-1} \left(\frac{x}{z}\right)^{n-2}.$$

and, by a similar reduction, there will be found,

$$1 = \frac{c_5}{a_5} \left(\frac{c}{a}\right)^{n-1} \left(\frac{x}{z}\right)^{n-2}.$$

When the relations which have been established are regarded in their application to the series of surfaces generated according to the law which has been assumed, the following conditions, as might have been anticipated, will be found to co-exist ; viz. :

$$\frac{c}{a} = \frac{c_2}{a_2} = \frac{c_4}{a_4} = \dots$$

$$\frac{c}{b} = \frac{c_2}{b_2} = \frac{c_4}{b_4} = \dots$$

$$\frac{b}{a} = \frac{b_2}{a_2} = \frac{b_4}{a_4} = \dots$$

$$\frac{c_l}{a_l} = \frac{c_3}{a_3} = \frac{c_5}{a_5} = \dots$$

$$\frac{c_l}{b_l} = \frac{c_3}{b_3} = \frac{c_5}{b_5} = \dots$$

$$\frac{b_l}{a_l} = \frac{b_3}{a_3} = \frac{b_5}{a_5} = \dots$$

It appears, therefore, that all the derived surfaces in which O Q, or p, meets at right angles the tangent planes, are similar to each other, and to the primary ; while those in which O P, or r, is perpendicular to the tangent planes are similar also to each other, and to the first derivative. The second set of surfaces are not, however, similar to the primitive, since,

$$\frac{c_l}{a_l} = \left(\frac{a}{c}\right)^{n-1}\left(\frac{z}{x}\right)^{n-2}$$

except only in the case of the ellipsoid, as was previously noticed in page 111.

(XXIV.) In the preceding section, the ratios of the axes have been established ; and it has been shewn that they are constant in alternating surfaces. It remains to ascertain the absolute values of these lines, as well as of the co-ordinates to the points in which the consecutive surfaces are intersected by the lines r or p ; the second investigation will be found to be implicitly involved in the former.

From page 116, (1), we have,

$$x_2 \;=\; \frac{u_i^2 \; u^{2(n-1)}}{a_i^n \; a^{n(n-1)}} \; x^{(n-1)^2}$$

$$\cdots \qquad\qquad \cdots$$

But, $$a_i \;=\; \frac{u^2 \, x^{n-2}}{a^{n-1}} \cdots ,$$

and, after the introduction of these equivalents, there will be found,

$$x_2 \;=\; \left(\frac{u_i}{u}\right)^2 x.$$

$$y_2 \;=\; \left(\frac{u_i}{u}\right)^2 y.$$

$$z_2 \;=\; \left(\frac{u_i}{u}\right)^2 z.$$

Now, since, $$\Sigma\left(\frac{x^n_2}{a^n_2}\right) \;=\; 1.$$

$$\therefore \; c_2{}^n \;=\; \left(\frac{c_2}{a_2}\right)^n x_2{}^n \;+\; \left(\frac{c_2}{b_2}\right)^n y_2{}^n \;+\; z_2{}^n.$$

$$= \; \left(\frac{c}{a}\right)^n\left(\frac{u_i}{u}\right)^{2n} x^n + \left(\frac{c}{b}\right)^n\left(\frac{u_i}{u}\right)^{2n} y^n + \left(\frac{u_i}{u}\right)^{2n} z^n.$$

$$= \; c^n \left(\frac{u_i}{u}\right)^{2n} \; \Sigma\left(\frac{x^n}{a^n}\right).$$

$$\therefore \; c_2{}^n \;=\; c^n \left(\frac{u_i}{u}\right)^{2n}.$$

consequently,

$$c_2 = \left(\frac{u_i}{u}\right)^2 c.$$

$$b_2 = \left(\frac{u_i}{u}\right)^2 b.$$

$$a_2 = \left(\frac{u_i}{u}\right)^2 a.$$

After a similar investigation, it will appear that,

$$x_3 = \left(\frac{u_2}{u_i}\right)^2 x_i. \qquad a_3 = \left(\frac{u_2}{u_i}\right)^2 a_i.$$

$$y_3 = \left(\frac{u_2}{u_i}\right)^2 y_i. \qquad b_3 = \left(\frac{u_2}{u_i}\right)^2 b_i$$

$$z_3 = \left(\frac{u_2}{u_i}\right)^2 z_i. \qquad c_3 = \left(\frac{u_2}{u_i}\right)^2 c_i.$$

The same principles being applicable to each of the consecutive surfaces, the following results may be tabulated.

$$x_i = \frac{u^2 x^{n-1}}{a^n}. \qquad y_i = \frac{u^2 y^{n-1}}{b^n}. \qquad z_i = \frac{u^2 z^{n-1}}{c^n}.$$

$$x_3 = \left(\frac{u_2}{u_i}\right)^2 x_i. \qquad y_3 = \left(\frac{u_2}{u_i}\right)^2 y_i. \qquad z_3 = \left(\frac{u_2}{u_i}\right)^2 z_i.$$

$$\cdots \qquad \cdots \qquad \cdots \qquad \cdots \qquad \cdots$$

$$x_2 = \left(\frac{u_i}{u}\right)^2 x. \qquad y_2 = \left(\frac{u_i}{u}\right)^2 y. \qquad z_2 = \left(\frac{u_i}{u}\right)^2 z.$$

$$x_4 = \left(\frac{u_3}{u_2}\right)^2 x_2. \qquad y_4 = \left(\frac{u_3}{u_2}\right)^2 y_2. \qquad z_4 = \left(\frac{u_3}{u_2}\right)^2 z_2.$$

$$\cdots \qquad \cdots \qquad \cdots \qquad \cdots$$

$$a_i = \frac{u^2 x^{n-2}}{a^{n-1}}. \qquad b_i = \frac{u^2 y^{n-2}}{b^{n-1}}. \qquad c_i = \frac{u^2 z^{n-2}}{c^{n-1}}.$$

$$a_3 = \left(\frac{u_2}{u_i}\right)^2 a_i. \qquad b_3 = \left(\frac{u_2}{u_i}\right)^2 b_i. \qquad c_3 = \left(\frac{u_2}{u_i}\right)^2 c_i.$$

$$\cdots \qquad \cdots \qquad \cdots \qquad \cdots \qquad \cdots$$

R

$$a_2 \;=\; \left(\frac{u_1}{u}\right)^2 a. \quad b_2 \;=\; \left(\frac{u_1}{u}\right)^2 b. \quad c_2 \;=\; \left(\frac{u_1}{u}\right)^2 c.$$

$$a_4 \;=\; \left(\frac{u_3}{u_2}\right)^2 a_2. \quad b_4 \;=\; \left(\frac{u_3}{u_2}\right)^2 b_2. \quad c_4 \;=\; \left(\frac{u_3}{u_2}\right)^2 c_2.$$

$$\ldots \qquad \ldots \qquad \ldots \qquad \ldots \qquad \ldots \qquad \ldots$$

(XXV.) For the calculation of the quantities $u\; u_1\; u_2\; \ldots$ the formula may be employed which is given in (9), page 25 (corrected); viz.:

$$u_m^{2n} \;=\; \frac{\Sigma^{n-1}\left\{ \dfrac{x^{\frac{m}{(n-1)}}}{a^{\frac{m}{n-1}}\frac{n(n-1)-2}{n-2}} \right\}^n}{\Sigma\left\{ \dfrac{x^{\frac{m+1}{(n-1)}}}{a^{\frac{m+1}{n-1}}\frac{n(u-1)-2}{n-2}} \right\}^n}.$$

This expression will be adapted to the consecutive surfaces which have been last considered, by writing 0, in the place of m; at the same time accenting the letters: so that,

$$u^{2n} \;=\; \frac{1}{\Sigma\left\{ \dfrac{x^{n-1}}{a^{n+1}} \right\}^n}.$$

$$u_i^{2n} \;=\; \frac{1}{\Sigma\left\{ \dfrac{x_i^{n-1}}{a_i^{n+1}} \right\}^n}.$$

The co-ordinates $x, y, z, \ldots$ as well as the semi-axes $a, b, c, \ldots$ are to be determined by means of the expressions given in (XXIV).

(XXVI.) In the two series of surfaces which have resulted from the preceding examination, the axes are proportional. For it may be seen that,

$$a_2 \ : \ b_2 \ : \ c_2 \ :: \ a \ : \ b \ : \ c.$$

$$a_4 \ : \ b_4 \ : \ c_4 \ :: \ a_2 \ : \ b_2 \ : \ c_2 \ :: \ a \ : \ b \ : \ c.$$

$$\ldots \qquad\qquad \ldots \quad \ldots \quad \ldots \quad \ldots \quad \ldots$$

$$a_3 \qquad b_3 \qquad c_3 \ :: \ a_{/} \ : \ b_{/} \ : \ c_{/}$$

$$a_5 \ : \ b_5 \ : \ c_5 \ :: \ a_3 \ : \ b_3 \ : \ c_3 \ :: \ a_{/} \ \cdot \ b_{/} \ : \ c_{/}$$

$$\ldots \quad \ldots \quad \ldots \quad \ldots \quad \ldots \qquad\qquad \ldots \quad \ldots$$

similarly, the co-ordinates to consecutive points of intersection in the two sets of surfaces are proportional, since it will appear that,

$$x_2 \ : \ y_2 \ : \ z_2 \ :: \ x \ : \ y \ : \ z.$$

$$x_4 \ : \ y_4 \ : \ z_4 \ :: \ x_2 \ : \ y_2 \ : \ z_2 \ :: \ x \ : \ y \ : \ z.$$

$$x_3 \ : \ y_3 \ : \ z_3 \ :: \ x_{/} \ : \ y_{/} \ : \ z_{/}$$

$$x_5 \ : \ y_5 \ : \ z_5 \ :: \ x_3 \ : \ y_3 \ : \ z_3 \ :: \ x_{/} \ : \ y_{/} \ : \ z_{/}.$$

$$\ldots \quad \ldots \quad \ldots \quad \ldots \quad \ldots \quad \ldots \quad \ldots$$

(XXVII.) Since, from page 121,

$$x_{/} \ = \ \frac{u^2 \, x^{n-1}}{a^n} \ ; \quad \text{and} \quad a_{/} \ = \ \frac{u^2 \, x^{n-2}}{a^{n-1}} \ ;$$

$$\frac{a x_{/}}{x} \ = \ \frac{u^2 \, x^{n-2}}{a^{n-1}} \ ;$$

$$a_{/} \ = \ \frac{x_{/}}{x} \, a : \quad \text{and,} \quad x_{/} \ = \ \frac{a_{/}}{a} \, x :$$

Again, by reference to the expressions which have been given in pages 121 and 122, for the successive axes and co-ordinates, there will be found,

$$\frac{x_2}{a_2} = \frac{x}{a} \; ; \quad \frac{x_3}{a_3} = \frac{x_i}{a_i} \; ; \quad \frac{x_4}{a_4} = \frac{x_2}{a_2} \; ; \quad \cdots$$

Consequently, the following series of relations will obtain,

$$x_i = \frac{a_i}{a} x \; ; \qquad y_i = \frac{b_i}{b} y \; ; \qquad z_i = \frac{c_i}{c} z \; ;$$

$$x_2 = \frac{a_2}{a} x \; ; \qquad y_2 = \frac{b_2}{b} y \; ; \qquad z_2 = \frac{c_2}{c} z \; ;$$

$$x_3 = \frac{a_3}{a_i} x_i \; ; \qquad y_3 = \frac{b_3}{b_i} y_i \; ; \qquad z_3 = \frac{c_3}{c_i} z_i \; ;$$

$$x_4 = \frac{a_4}{a_2} x_2 \; ; \qquad y_4 = \frac{b_4}{b_2} y_2 \; ; \qquad z_4 = \frac{c_4}{c_2} z_2 \; ;$$

$$\cdots \qquad \cdots \qquad \cdots \qquad \cdots \qquad \cdots \qquad \cdots$$

which may be immediately reduced to the form,

$$x_i = \frac{a_i}{a} x \; ; \qquad y_i = \frac{b_i}{b} y \; ; \qquad z_i = \frac{c_i}{c} z \; ;$$

$$x_2 = \frac{a_2}{a} x \; ; \qquad y_2 = \frac{b_2}{b} y \; ; \qquad z_2 = \frac{c_2}{c} z \; ;$$

$$x_3 = \frac{a_3}{a} x \; ; \qquad y_3 = \frac{b_3}{b} y \; ; \qquad z_3 = \frac{c_3}{c} z \; ;$$

$$\cdots \qquad \cdots \qquad \cdots \qquad \cdots \qquad \cdots \qquad \cdots$$

$$x_m = \frac{a_m}{a} x. \qquad y_m = \frac{b_m}{b} y. \qquad z_m = \frac{c_m}{c} z.$$

(XXVIII.) By combining the expressions deduced in the preceding page the following remarkable singularity may be demonstrated.

Since,
$$x_, x_2 \;=\; a_, a_2 \left(\frac{x^2}{a^2}\right);$$

$$\therefore \;\; \left(\frac{x_, x_2}{a_, a_2}\right)^{\frac{1}{2}} \;=\; \frac{x}{a}:$$

consequently,
$$\Sigma \left(\frac{x_, x_2}{a_, a_2}\right)^{\frac{n}{2}} \;=\; 1.$$

In the same way, it will appear that,

$$\Sigma \left(\frac{x_, x_2 \, x_3}{a_, a_2 \, a_3}\right)^{\frac{n}{3}} \;=\; 1; \qquad \Sigma \left(\frac{x_, x_2 \, x_3 \, x_4}{a_, a_2 \, a_3 \, a_4}\right)^{\frac{n}{4}} \;=\; 1;$$

and, generally,

$$\Sigma \left(\frac{x_, x_2 \, x_3 \, \ldots \, x_m}{a_, a_2 \, a_3 \, \ldots \, a_m}\right)^{\frac{n}{m}} \;=\; 1.$$

This relation may be otherwise shown in the following manner,

$$\Sigma \left(\frac{x}{a}\right)^n \;=\; 1;$$

and, $\left(\dfrac{x}{a}\right)^n = \left(\dfrac{x^m}{a^m}\right)^{\frac{n}{m}} = \left(\dfrac{x}{a}\,\dfrac{x}{a}\,\ldots\,\dfrac{x}{a}\right)^{\frac{n}{m}} = \left(\dfrac{x_,}{a_,}\,\dfrac{x_2}{a_2}\,\ldots\,\dfrac{x_m}{a_m}\right)^{\frac{n}{m}}$

$\therefore \;\; \left(\dfrac{x_, x_2 \ldots x_m}{a_, a_2 \ldots a_m}\right)^{\frac{n}{m}} + \left(\dfrac{y_, y_2 \ldots y_m}{b_, b_2 \ldots b_m}\right)^{\frac{n}{m}} + \left(\dfrac{z_, z_2 \ldots z_m}{c_, c_2 \ldots c_m}\right)^{\frac{n}{m}} = 1.$

(XXIX.) As a certain degree of intricacy attaches to many of the previous investigations, it may be desirable to give, in conclusion, some verifications of the results which have been deduced for the general surface, by reference to the particular case of the ellipsoid; since the corresponding expressions applying to that surface have been obtained independently.

In order to ascertain the ratio $\frac{u_{\prime}}{u}$, we find from (3), page 17, the fundamental equation,

$$\frac{1}{u_m{}^{2n}} \quad = \quad \Sigma \left\{ \frac{x_m{}^{n(n-1)}}{a_m{}^{n(n+1)}} \right\}.$$

consequently, $\quad \dfrac{1}{u^{2n}} = \Sigma \left\{ \dfrac{x^{n(n-1)}}{a^{n(n+1)}} \right\}. \quad \dfrac{1}{u_{\prime}{}^{2n}} = \Sigma \left\{ \dfrac{x_{\prime}{}^{n(n-1)}}{a_{\prime}{}^{n(n+1)}} \right\}.$

But it has been shown, in page 111, (XVII), that,

$$x_{\prime} \quad = \quad \frac{u^2 \, x^{n-1}}{a^n}; \qquad a_{\prime} \quad = \quad \frac{u^2 \, x^{n-2}}{a^{n-1}};$$

and from these relations it will be found that,

$$\Sigma \left\{ \frac{x_{\prime}{}^{n(n-1)}}{a_{\prime}{}^{n(n+1)}} \right\} \quad = \quad \frac{1}{u^{4n}} \, \Sigma \left\{ \frac{a^{n(n-1)}}{x^{n(n-3)}} \right\}.$$

$$\therefore \quad \frac{u^{4n}}{u_{\prime}{}^{2n}} \quad = \quad \Sigma \left\{ \frac{a^{u(n-1)}}{x^{n(n-3)}} \right\}.$$

Now, when the general surface is reduced to the ellipsoid, by the assumption $n = 2$, this expression takes the form,

$$\left(\frac{u^2}{u_{\prime}} \right)^4 \quad = \quad \Sigma \, (a^2 \, x^2).$$

But, in page 102, it has been assumed that,

$$v^4 \quad = \quad \Sigma \, (a^2 \, x^2);$$

consequently, $\qquad \dfrac{u_{\prime}}{u} \quad = \quad \dfrac{u}{v}.$

In the general surface of the nth degree it has been shewn, in page 120, that,

$$v_2 \quad = \quad \left(\frac{u_{\prime}}{u} \right)^2 x; \qquad a_2 \quad = \quad \left(\frac{u_{\prime}}{u} \right)^2 a;$$

In the ellipsoid, therefore, these formulæ will be reduced to the expressions,

$$x_2 = \left(\frac{u}{v}\right)^2 x. \qquad a_2 = \left(\frac{u}{v}\right)^2 a.$$

These relations will be verified by referring to the results recorded in (VII), page 103 ; and in (VIII), page 105.

Again, if we wish to verify the expressions which have been given for x_3 ... , there will be found, after the requisite reductions,

$$\left(\frac{u_2}{u_{\prime}}\right)^{2n} = \left(\frac{u_{\prime}^2}{u^3}\right)^{2n} \Sigma\left\{\frac{a^{n(n-1)}}{x^{n(n-3)}}\right\}.$$

Consequently, if $n = 2$,

$$\left(\frac{v_2}{u_{\prime}}\right)^4 = \left(\frac{u_{\prime}^2}{u^3}\right)^4 . \ \Sigma\left\{a^2\ x^2\right\}$$

$$= \left(\frac{u_{\prime}^2}{u^3}\right)^4 v^4.$$

$$\therefore \quad \frac{u_2}{u_{\prime}} = \frac{u_{\prime}^2}{u^3} v. = \left(\frac{u_{\prime}}{u}\right)^2\left(\frac{v}{u}\right).$$

But, from page 126, $\quad \dfrac{u_{\prime}}{u} = \dfrac{u}{v}$;

consequently, $\qquad x_3 = \left(\dfrac{u}{v}\right)^2 x :$

and, $\qquad a_3 = \left(\dfrac{u^2}{v}\right)^2 \dfrac{1}{a} :$

a result which agrees perfectly with that which was demonstrated in the case of the ellipsoid, in page 104. These verifications, by reference to particular instances which may be multiplied to any extent, appear to suffice in confirmation of the formulæ which belong to the general surface of the nth degree.

(XXX.)   It will now be shewn that the formula for $u_m{}^{2n}$, as given in page 122, or 25 (corrected), identifies itself with the fundamental expression (3), in page 17.

Let $m = 0$, then $\Sigma^{n-1}\left(\dfrac{x}{a}\right)^n = 1$; and, after inverting the terms of the formula in page 122, we shall obtain,

$$\frac{1}{u^{2n}} = \Sigma\left\{\frac{x^{n(n-1)}}{a^{n(n+1)}}\right\}.$$

which agrees with (3) as given at page 17.

Again, if $m = 1$, the index of a in the denominator of the general form in page 122, will be $\dfrac{n(n-1)^2-2}{n-2} = n^2 + 1$;

consequently, the expression reduces itself to,

$$u_{,}{}^{2n} = \frac{\Sigma^{n-1}\left\{\dfrac{x^{n(n-1)}}{a^{n(n+1)}}\right\}}{\Sigma\left\{\dfrac{x^{\frac{n(n-1)}{2}}}{a^{n(n+1)}}\right\}} = \frac{1}{u^{2n(n-1)}}\cdot\frac{1}{\Sigma\left\{\dfrac{x^{\frac{n(n-1)}{2}}}{a^{n(n+1)}}\right\}}.$$

$$\therefore \quad \frac{1}{u_{,}{}^{2n}\,u^{2n(n-1)}} = \Sigma\left\{\frac{x^{\frac{n(n-1)}{2}}}{a^{n(n+1)}}\right\}.$$

But the formula at page 17 becomes, when $m = 1$,

$$\frac{1}{u_{,}{}^{2n}} = \Sigma\left\{\frac{x_{,}{}^{n(n-1)}}{a^{n(n+1)}}\right\};$$

in which, $x_{,} = \dfrac{n^2\,x^{n-1}}{a^n}.$

The substitution for $x_{,}$ of its equivalent value, will yield an expression identical with that which has just been deduced from the form given in (XXV).

(XXXI.)   In employing the general formula of (XXV), page 122, m may be assumed to have any positive integral values, as in the last section, under the restriction that the lines are confined to a single surface,—e. g., the primary; or, otherwise, that they belong to consecutive surfaces passing through points determined upon the primary by the law of tangential intersection described in (III), page 16.

When, however, this formula is applied to consecutive surfaces which are generated according to the hypothesis in (XX), page 115, we have to write $m = 0$ in the indices, at the same time replacing $x\,y\,z$ by $x_i\,y_i\,z_i$ ..., $x_m\,y_m\,z_m$. Then, for the mth surface,

$$\Sigma^{n-1}\left\{ \frac{x^{\frac{m}{(n-1)}}}{a^{\frac{m}{n-2}}} \right\}^n = \Sigma^{n-1}\left\{ \frac{x_m^{(n-1)}}{a_m^{n-2}} \right\}^n = \Sigma^{n-1}\left\{ \frac{x_m^n}{a_m^n} \right\} = 1.$$

$$\Sigma\left\{ \frac{x^{\frac{m+1}{(n-1)}}}{a^{\frac{m+1}{n-2}}} \right\}^n = \Sigma\left\{ \frac{x_m^{n-1}}{a_m^{\frac{n(n-1)-2}{n-2}}} \right\}^n = \Sigma\left\{ \frac{x_m^{n(n-1)}}{a_m^{n(n+1)}} \right\}.$$

$$\therefore \quad \frac{1}{u_m^{2n}} = \Sigma\left\{ \frac{x_m^{n(n-1)}}{a_m^{n(n+1)}} \right\}. \tag{1}$$

Now, in the formula (3), page 17, the value of u_m has been deduced for consecutive lines in the same surface; and, in order that it may be applicable to lines in the consecutive derivatives which are now under consideration, a_m is to be substituted for a: the expression is then, consistently, coincident with (1).

Cor.   It is evident that the results obtained from the value of u_m in page 122, ought to be identical with those which issue from the general form (3), in page 23.   In illustration of this, let $m = 2$ in that formula; the result will be,

s

$$\Sigma \left\{ \frac{x^{\frac{(n-1)^2}{2}}}{a^{\frac{n(n-1)-2}{n-2}}} \right\}^n \;=\; \frac{1}{(p, r_2)^n (p \; r)^{n(n-1)}} :$$

$$\therefore \;\; \frac{1}{u_{\prime}^{2n} u^{2n(n-1)}} \;=\; \Sigma \left\{ \frac{x^{\frac{n(n-1)^2}{2}}}{a^{n(n+1)}} \right\}.$$

which agrees with the expression obtained in page 128 by writing $m = 1$.

(XXXII.) If any three consecutive surfaces are considered in the series generated according to the hypothesis in (XX), page 115; and if $u_{m-1} \; u_m \; u_{m+1}$ are the circle-ordinates connected with the three surfaces respectively; these lines are always subject to the relation,

$$u^2{}_m \;=\; u_{m-1} \, u_{m+1}.$$

From (1), (XXXI),

$$\frac{1}{u^{2n}{}_{m+1}} \;=\; \Sigma \left\{ \frac{x^{n(n-1)}_{m+1}}{a^{n(n+1)}_{m+1}} \right\}.$$

But, from page 121,

$$x_{m+1} \;=\; \left(\frac{u_m}{u_{m-1}} \right)^2 x_{m-1} : \qquad a_{m+1} \;=\; \left(\frac{u_m}{u_{m-1}} \right)^2 a_{m-1} :$$

consequently,

$$\frac{x^{n(n-1)}_{m+1}}{a^{n(n+1)}_{m+1}} \;=\; \frac{u^{4n}_{m-1}}{u^{4n}_m} \frac{x^{n(n-1)}_{m-1}}{a^{n(n+1)}_{m-1}}.$$

Hence, we obtain,

$$\frac{1}{u^{2n}{}_{m+1}} \;=\; \frac{u^{4n}{}_{m-1}}{u^{4n}{}_{m}} \; \Sigma \left\{ \frac{x^{n(n-1)}{}_{m-1}}{a^{n(n+1)}{}_{m-1}} \right\} \;=\; \frac{u^{4n}{}_{m-1}}{u^{4n}{}_{m}} \; \frac{1}{u^{2n}{}_{m-1}}$$

$$\therefore \quad u^{4n}{}_{m} \;=\; u^{2n}{}_{m-1} \; u^{2n}{}_{m}+1$$

$$\therefore \quad u^{2}{}_{m} \;=\; u_{m-1} \; u_{m}+1.$$

From this proposition, then, we obtain the singular property, that the circle-ordinate in any of these derived surfaces is a mean proportional to those which belong to the two surfaces immediately preceding and following in the series; so that, when consecutive values are assigned to m,

$$u_1^{2} \;=\; u \; u_2$$

$$u_2^{2} \;=\; u_1 \; u_3$$

$$u_3^{2} \;=\; u_2 \; u_4$$

$$\cdots \qquad \cdots$$

Cor. 1.—When the surfaces are ellipsoids, we find, by employing the formulæ of (VII) (VIII), pages 103—106, the curious series of relations following :

$$u^{2} \;=\; u_1 \; v.$$

$$u^{3} \;=\; u_2 \; v^{2}.$$

$$u^{4} \;=\; u_3 \; v^{3}.$$

$$\cdots \qquad \cdots$$

$$u^{m} \;=\; u_{m-1} \; v^{m-1}.$$

The elimination of v from these equations will reproduce the expressions which have been determined, connecting the successive values of u_m.

Cor. 2.—Since $x_m = \dfrac{a_m}{a} x$, page 124, the value of u_m may be expressed in terms of the axes of its proper surface. Thus, we find,

$$\frac{1}{u_m^{2n}} \;=\; \Sigma \left\{ \frac{1}{a_m^{2n}} \left(\frac{x}{a} \right)^{n(n-1)} \right\} .$$

Cor. 3.—Hence, in the general surface,

$$\Sigma \left(\frac{1}{u^{2n}} \right) = \left(\frac{x}{a} \right)^{n(n-1)} \Sigma \left(\frac{1}{a^{2n}} \right) + \left(\frac{y}{b} \right)^{n(n-1)} \Sigma \left(\frac{1}{b^{2n}} \right) + \left(\frac{z}{c} \right)^{n(n-1)} \Sigma \left(\frac{1}{c^{2n}} \right).$$

This includes the relation in page 102, which has been established independently in reference to the ellipsoid.

CHAPTER VI.

(I.) As a question connected with the main subject of this volume, it is now proposed to determine the area of a section of the ellipsoid, formed by a plane cutting the surface in any manner whatever.

In the particular case when the plane is drawn through the centre, the expression for a sectional area is well known, and will be found in most works which treat of Solid Geometry; but the author is not aware that any solution has hitherto been given to this problem in its general sense.

Since a determination of the area will lead readily to that of the volume of any segment or frustum, this investigation, independently of such elegance as may attach to the expression deduced, appears to be not altogether devoid of a practical interest: while the value of the results seems to be still further enhanced, in regarding the facility with which, as will be seen, they may be made to bear upon the solution of many questions relating to the Attraction of the Solid.

Let A be the area of any section, which can be shewn to be, generally, elliptic.

p a perpendicular drawn from the centre to the secant plane.

p_i a perpendicular upon the tangent plane which is parallel to, and least remote from, the secant plane.

The expression representing the area of a section will then be found to be,

$$A = \frac{\pi a b c}{p_i^3} \left(p_i^2 - p^2 \right) :$$

in which $a\,b\,c$ are the semiaxes of the surface.

(II.)   Let the equation of a plane cutting the ellipsoid in any manner be,

$$z \;=\; mx \;+\; ny \;+\; e\,.$$

By eliminating z, between this equation and that of the surface, viz., $\Sigma\!\left(\dfrac{x^2}{a^2}\right) \;=\; 1$, we obtain the equation of a vertical elliptic cylinder which includes the section; or, which amounts to the same thing, an equation to the orthographical projection representing the sectional curve upon the plane of xy.

There is, then,

$$\frac{x^2}{a^2} \;+\; \frac{y^2}{b^2} \;+\; \frac{(mx + ny + e)^2}{c^2} \;=\; 1:$$

from which will be obtained, as the equation of the elliptic projection,

$$(n^2 b^2 \;+\; c^2)\, a^2 y^2 \;+\; (m^2 a^2 \;+\; c^2)\, b^2 x^2 \;+\; 2mn\, a^2 b^2\, xy$$

$$+\; 2m\, a^2 b^2\, e\, x \;+\; 2n\, a^2 b^2\, e\, y \;+\; a^2 b^2 (e^2 - c^2) \;=\; 0. \qquad (1)$$

(III.)   Let h and k be co-ordinates to the centre of this projection (1); $x_{\prime}, y_{\prime}$ those of the curve measured from its centre: so that,

$$x \;=\; x_{\prime} \;+\; h\,. \qquad y \;=\; y_{\prime} \;+\; k\,.$$

By introducing the new co-ordinates in (1), the constant term being denoted by $\phi(h\,k)$, there will be found,

$$(n^2 b^2 \;+\; c^2)\, a^2 y_{\prime}^2 \;+\; (m^2 a^2 \;+\; c^2)\, b^2 x_{\prime}^2 \;+\; 2mn\, a^2 b^2\, x_{\prime} y_{\prime}$$

$$+\; \phi(h\,k) \;=\; 0; \qquad (2)$$

the equations of condition being given, in the usual manner, by assuming the co-efficients of $x_{\prime}$ and $y_{\prime}$ respectively $= 0$: then,

$$(n^2 b^2 \;+\; c^2)\, a^2 k \;+\; mn\, a^2 b^2\, h \;+\; n\, a^2 b^2\, e \;=\; 0. \qquad (3)$$

$$(m^2 a^2 \;+\; c^2)\, b^2 h \;+\; mn\, a^2 b^2\, k \;+\; m\, a^2 b^2\, e \;=\; 0. \qquad (4)$$

From (3) (4) the co-ordinates of the centre are ascertained in the terms following ;

$$h = - \frac{m \, a^2 \, e}{m^2 \, a^2 + n^2 \, b^2 + c^2}.$$

$$k = - \frac{n \, b^2 \, e}{m^2 \, a^2 + n^2 \, b^2 + c^2}.$$

There is, further, the constant term,

$$\phi \, (h \, k) = (n^2 \, b^2 + c^2) \, a^2 \, k^2 + (m^2 \, a^2 + c^2) \, b^2 \, h^2 + 2m \, n \, a^2 \, b^2 \, h \, k$$

$$+ \, 2m \, a^2 \, b^2 \, e \, h + 2n \, a^2 \, b^2 \, e \, k + a^2 \, b^2 \, (e^2 - c^2) \, ;$$

and from this will result, after the requisite reductions,

$$\phi \, (h \, k) = a^2 \, b^2 \, c^2 \left(\frac{e^2}{m^2 \, a^2 + n^2 \, b^2 + c^2} - 1 \right). \tag{6}$$

(IV.) In order to ascertain the position and dimensions of the principal axes in this projection, let $x^2 \, y^2$ be the rectangular co-ordinates of the curve referred to those lines ; θ the angle at which the axis major is inclined to the present axis of x or $x_{,}$: we have then, by the ordinary method of transformation,

$$x_{,} = x_2 \cos \theta - y_2 \sin \theta.$$

$$y_{,} = x_2 \sin \theta + y_2 \cos \theta.$$

By the introduction of these expressions, equation (2) will be changed into,

$$\left\{ (n^2 b^2 + c^2) \, a^2 \sin {}^2\theta + (m^2 a^2 + c^2) \, b^2 \cos {}^2\theta + m \, n \, a^2 b^2 \sin 2\theta \right\} x^2{}_2 \; +$$

$$\left\{ (n^2 b^2 + c^2) \, a^2 \cos {}^2\theta + (m^2 a^2 + c^2) \, b^2 \sin {}^2\theta - m \, n \, a^2 b^2 \sin 2\theta \right\} y^2{}_2 \; +$$

$$\left\{ (n^2 b^2 + c^2) \, a^2 \sin 2\theta - (m^2 a^2 + c^2) \, b^2 \sin 2\theta + 2m \, n \, a^2 b^2 \cos 2\theta \right\} \; x_2 \, y_2$$

$$+ \quad \phi \, (h \, k) \quad = \quad 0. \tag{7}$$

When the co-efficient of $x_2\, y_2$ is equated to zero, there will be,

$$\tan 2\theta = \frac{2m\,n\,a^2\,b^2}{(m^2\,a^2 + c^2)\,b^2 - (n^2\,b^2 + c^2)\,a^2}.\qquad(8)$$

From (8) the principal axes become known in *position ;* and the equation of the projected curve, referred to these axes, may be written,

$$A_2\,y^2_2 \;+\; B_2\,x^2_2 \;=\; f:\qquad(9)$$

in which, $\quad f \;=\; -\,\phi\,(h\,k) \;=\; a^2\,b^2\,c^2\!\left(1 - \dfrac{e^2}{m^2\,a^2 + n^2\,b^2 + c^2}\right).$

(V.) For determining the *dimensions* of the principal axes, we have from (7),

$$A_2 \;=\; \Big\{(n^2\,b^2 + c^2)a^2 \cos^2\theta + (m^2\,a^2 + c^2)\,b^2 \sin^2\theta - m\,n\,a^2\,b^2 \sin 2\theta\Big\}$$

$$\therefore\; 2A_2 \;=\; (m^2\,a^2 + c^2)\,b^2 + (n^2\,b^2 + c^2)\,a^2 -$$

$$\Big\{(m^2\,a^2 + c^2)\,b^2 - (n^2\,b^2 + c^2)\,a^2\Big\}\,\cos 2\theta - 2m\,n\,a^2\,b^2 \sin 2\theta.$$

But, from (8), it will appear that,

$$\Big\{(m^2\,a^2 + c^2)\,b^2 - (n^2\,b^2 + c^2)\,a^2\Big\}\,\cos 2\theta + 2m\,n\,a^2\,b^2 \sin 2\theta$$

$$=\; 2m\,n\,a^2\,b^2\,\operatorname{cosec} 2\theta\,;$$

consequently,

$$A_2 \;=\; \frac{1}{2}\Big\{(m^2\,a^2 + c^2)\,b^2 + (n^2\,b^2 + c^2)\,a^2 - 2m\,n\,a^2\,b^2\,\operatorname{cosec} 2\theta\Big\}\,.$$

Similarly,

$$B_2 \;=\; \frac{1}{2}\Big\{(m^2\,a^2 + c^2)\,b^2 + (n^2\,b^2 + c^2)\,a^2 + 2m\,n\,a^2\,b^2\,\operatorname{cosec} 2\theta\Big\}\,.$$

If $a_2\,b_2$ are the semi-axes of the projection, it will be represented by,

$$\frac{y_2^{\,2}}{b_2^{\,2}} \;+\; \frac{x_2^{\,2}}{a_2^{\,2}} \;=\; 1\,;\qquad(10)$$

and, by comparing this equation with (2), we shall have, finally,

$$a_2 = \frac{(2f)^{\frac{1}{2}}}{\left\{ (m^2\, a^2 + c^2)\, b^2 + (n^2\, b^2 + c^2)\, a^2 + 2m\, n\, a^2\, b^2\, \operatorname{cosec}\, 2\theta \right\}^{\frac{1}{2}}}.$$

$$b_2 = \frac{(2f)^{\frac{1}{2}}}{\left\{ (m^2\, a^2 + c^2)\, b^2 + (n^2\, b^2 + c^2)\, a^2 - 2m\, n\, a^2\, b^2\, \operatorname{cosec}\, 2\theta \right\}^{\frac{1}{2}}}.$$

(VI.)   Now, if Λ' is the area of the projected ellipse,

$$A' = \pi\, a_2\, b_2 :$$

in which expression,

$$a_2\, b_2 = \frac{2f}{\left\{ \left\{ (m^2\, a^2 + c^2)\, b^2 + (n^2\, b^2 + c^2)\, a^2 \right\}^2 - 4m^2\, n^2\, a^4 b^4\, \operatorname{cosec}^2\, 2\theta \right\}^{\frac{1}{2}}}.$$

Let D be written for the denominator of this fraction; then, after eliminating θ by means of the equation (8), we shall obtain,

$$D = 2\, a\, b\, c\, (m^2\, a^2 + n^2\, b^2 + c^2)^{\frac{1}{2}}.$$

When the value of f has been introduced from page 136, the area of projection will be ascertained under the following expression,

$$\Lambda' = \frac{\pi\, a\, b\, c}{m^2\, a^2 + n^2\, b^2 + c^2)^{\frac{1}{2}}} \left(1 - \frac{e^2}{m^2\, a^2 + n^2\, b^2 + c^2} \right). \quad (11)$$

(VII.)   In order to simplify the equation (11), let $x_i\, y_i\, z_i$ be co-ordinates of that point in which the ellipsoid is in contact with a tangent plane drawn parallel to the secant, the equations of the two planes will be,

$$\frac{x\, x_i}{a^2} + \frac{y\, y_i}{b^2} + \frac{z\, z_i}{c^2} = 1 :$$

$$z = m\, x + n\, y + e :$$

T

and by combining these expressions, under the condition of parallelism, there will ensue, for contact,

$$z_{,} = \pm \frac{c^2}{(m^2 a^2 + n^2 b^2 + c^2)^{\frac{1}{2}}}.$$

$$y_{,} = \mp \frac{nb^2}{(m^2 a^2 + n^2 b^2 + c^2)^{\frac{1}{2}}}.$$

$$x_{,} = \mp \frac{ma^2}{(m^2 a^2 + n^2 b^2 + c^2)^{\frac{1}{2}}}.$$

Now, since, $\dfrac{1}{p_{,}^{2}} = \Sigma\left(\dfrac{x_{,}^{2}}{a^4}\right)$, the foregoing values will give,

$$\frac{1}{p_{,}} = \left(\frac{m^2 + n^2 + 1}{m^2 a^2 + n^2 b^2 + c^2}\right)^{\frac{1}{2}}. \tag{12}$$

Again, if i is the inclination of the secant, or tangent plane, to that of $x\,y$,

$$\sec i = (m^2 + n^2 + 1)^{\frac{1}{2}} :$$

consequently, $p_{,} \sec i = (m^2 a^2 + n^2 b^2 + c^2)^{\frac{1}{2}}.$

We have, also, $p = e \cos i$;

$$\therefore \; (m^2 a^2 + n^2 b^2 + c^2)^{\frac{1}{2}} = \frac{e\,p_{,}}{p} :$$

and, $$\frac{e^2}{m^2 a^2 + n^2 b^2 + c^2} = \frac{p^2}{p_{,}^{2}}.$$

Hence, $\therefore$ $$A' = \frac{\pi\,abc}{p_{,}^{3}} (p_{,}^{2} - p^2) \frac{p}{e}. \tag{13}$$

(VIII.)   Since A' is the orthographical projection of the given elliptic section A, there is,

$$A = A' \sec i = A' \frac{e}{p}$$

$$\therefore \; A = \frac{\pi\,abc}{p_{,}^{3}} (p_{,}^{2} - p^2). \tag{14}$$

COR. 1.—If $p = 0$, the secant plane will pass through the centre; let this section be called A_i : the general expression then takes the well-known form,

$$A_i = \frac{\pi a b c}{p_i}.$$

COR. 2.—Consequently,

$$A = A_i \left(1 - \frac{p^2}{p_i^2}\right).$$

(VIII.) The relation deduced in Cor. 2 of the preceding section suggests an elegant geometrical interpretation of the formula (14), which expresses generally a sectional area of the ellipsoid.

In figure 4, let O P, a vectorial radius to the point of contact P, cut the secant plane in k.

Suppose O n to be drawn from the centre of the surface, perpendicular to the tangent-plane at P ; and cutting the sectional plane in the point m.

Let O n be produced to meet in the point n_i the parallel tangent plane, on the opposite side of the surface ; and, upon $n n_i$ as diameter suppose a circle to be described, in the plane O P n, which contains the vectorial radius O P together with the tangent-perpendicular O n.

Produce $k m$, in the plane O P n, to meet this circle in Q.

Join O Q, and assume the angle Q O $n = \theta$.

From this construction we obtain, $p = p_i \cos \theta$;

$$\therefore \quad 1 - \frac{p^2}{p_i^2} = 1 - \left(\frac{O m}{O Q}\right)^2 = 1 - \cos^2 \theta.$$

$$\therefore \quad A = A_i \sin^2 \theta = A_i \cos\left(\frac{\pi}{2} - \theta\right) \times \cos\left(\frac{\pi}{2} - \theta\right).$$

Now, let A_2 be the orthographic projection of A_1, the central section, upon a plane which contains the line OQ, at the same time that it cuts at right angles the primary plane QOP.

The inclination of A_1 to this new plane is the angle QOL, $= \frac{\pi}{2} - \theta$;

$$\therefore \quad A_2 = A_1 \cos\left(\frac{\pi}{2} - \theta\right) :$$

consequently, $\qquad A = A_2 \cos\left(\frac{\pi}{2} - \theta\right)$.

Finally, let A_3 be the area of A_2, when re-projected upon the original central plane section A_1 ; i.e., the plane LL_1 in the figure 4 : then,

$$A = A_2 \cos\left(\frac{\pi}{2} - \theta\right) = A_3.$$

(IX.) The considerations adduced in the preceding article lead to the singular property of the ellipsoid which is included in the following proposition.

THEOREM.—The area of any section formed by a plane cutting the ellipsoid is equal to the second projection of the parallel central section. The projection being made, orthographically ;—1st, upon a plane determined in position from that of the original section : 2nd, by the re-projection of the first projected area upon the plane of the primary central section.

(X.) From the formula which has been given in (14), for the area of a section of the ellipsoid, may be derived a general expression for the VOLUME of any portion of the surface, limited by parallel planes.

If V represents the volume contained in any segment of an ellipsoid, of which the base is A, there will be ultimately the differential equation,

$$dV = A\,dp.$$

$$\therefore \quad dV = \frac{\pi\,abc}{p_1^3}\,(p_1^2 - p^2)\,dp.$$

The integration of this expression will give the volume of any portion of the solid which is bounded by parallel planes; viz. :

$$V + C = \frac{\pi\,abc}{3\,p_i^{3}}\,(3\,p_i^{2}\,p - p^{3}). \tag{1}$$

Cor. 1.—When this integral has been corrected between the limits — p_i and + p_i, there will be found, for the volume of the whole solid, the usual expression,

$$V = \frac{4\,\pi abc}{3}.$$

Cor. 2.—In the formula (1), V may be regarded as a function of θ, from the relation in (VIII) $p = p_i \cos\theta$; then,

$$V + C = \frac{\pi\,abc}{3}\,(3\cos\theta - \cos^{3}\theta) :$$

the limits of θ being π and 0, for the whole solid.

Cor. 3.—To determine the volume of any segment or frustum of the ellipsoid.

Let p_2 be the distance of any plane section from the centre. The formula (1) will then give,

$$C = \frac{\pi\,abc}{3\,p_i^{3}}\,(3\,p_i^{2}\,p_2 - p^{3}_2)$$

$$\therefore\quad V = \frac{\pi\,abc}{3\,p_i^{3}}\left\{3\,p_i^{2} - (p^{2} + p\,p_2 + p^{2}_2)\right\}(p - p_2): \tag{2}$$

expressing the volume of any *frustum*, of which the limiting planes are at distances p_2 and p from the centre.

If, in (2), $p = p_i$, we find for the *segment*,

$$V = \frac{\pi\,abc}{3\,p_i^{3}}\,(p_i - p_2)^{2}\,(2\,p_i + p_2). \tag{3}$$

p_2 being the perpendicular drawn to the base of the segment.

Cor. 4.—In (1) or (3) suppose p or p_{2} to be perpendicular to the tangent plane of a similar concentric ellipsoid ; then $\dfrac{p}{p_{i}}$ or $\dfrac{p_{2}}{p_{i}}$ is constant.

Hence is evident the well-known property that the volume is constant intercepted by a plane touching the ellipsoid, and cutting an external similar surface.

(XI.) In this section we adduce examples in illustration of the formulæ which have been demonstrated, in reference to the volume of an ellipsoid.

PROBLEM 1.—It is required to *trisect* the ellipsoid.

If p and p_{2} are perpendiculars drawn from the centre upon parallel planes which limit the lateral dimensions of any frustum of the solid, p_{i} being a perpendicular upon the tangent plane which is parallel to the former ; we have, from (X), (2) ;

$$V_{i} \;=\; \frac{\pi a b c}{3\, p_{i}^{3}} \left\{ 3\, p_{i}^{2} - (p^{2} + p\, p_{2} + p_{2}^{2}) \right\} (p-p_{2}). \qquad (1)$$

In the case under consideration the terminal segments are equal, and each equal to the frustum, which will be symmetrically related to the centre. In order to express the last condition analytically, we have to write $p_{2} = -\,p$; then,

$$V_{i} \;=\; \frac{2\,\pi a b c}{3\, p_{i}^{3}} \, (3\, p_{i}^{2} - p^{2})\, p.$$

From (X), (3), the volume of a segment, in relation to which p_{2} is a perpendicular drawn from the centre to the limiting plane, is,

$$V_{2} \;=\; \frac{\pi a b c}{3\, p_{i}^{3}} \, (p_{i} - p_{2})^{2} \, (2\, p_{i} + p_{2}) ;$$

and this will be contiguous to the former section if we make $p_{2} = p$: then,

$$V_{2} \;=\; \frac{\pi a b c}{3\, p_{i}^{3}} \, (p_{i} - p)^{2} \, (2\, p_{i} + p)$$

Equating V_i and V_2 we find,

$$3 p^3 \;-\; 9 p_i^2 p \;+\; 2 p_i^3 \;=\; 0. \tag{2}$$

The solution of this cubic will give the value of p required.

Since $p = p_i \cos \theta$, let $\cos \theta = v$;

$$\therefore\; 3 v^3 \;-\; 9 v \;+\; 2 \;=\; 0.$$

On applying Sturm's Theorem we find that there are, analytically, three real roots, two of which are excluded, as being beyond the possible limits of $\cos \theta$; and, by employing the trigonometrical method of solution, there will be obtained, as the only admissible value,

$$v \;=\; 0.2261 \;=\; \cos \theta.$$

$$\therefore\; p \;=\; 0.2261\, p_i.$$

This determines the bounding planes of the two segments, the thickness of the frustum being, consequently,

$$2 p \;=\; 0.4522\, p_i.$$

Cor. 1.—It is to be remarked as singular that the value of p is always the same fraction of p_i, in whatever position the parallel trisecting planes are taken; in this respect, therefore, the property of the sphere is retained in the general surface.

Cor. 2.—If the ellipsoid is reduced to a sphere, p_i is constant for all positions, and $= a$ the radius. Let x be an abscissa measured from the centre; then, from the usual formula $V = \pi \int y^2\, dx$, there is,

$$V_2 \;=\; \frac{2\pi a^3}{3} \;-\; \pi\left(a^2 x \;-\; \frac{x^3}{3}\right).$$

$$=\; \frac{\pi}{3}\, (a - x)^2\, (2a + x).$$

$$V_i \;=\; 2\pi\left(a^2 x \;-\; \frac{x^3}{3}\right).$$

After equating the foregoing expressions, we shall find,

$$3x^3 \ - \ 9a^2 x \ + \ 2a^3 \ = \ 0 \ ;$$

an equation which coincides with (2), (XI), obtained from the ellipsoid, and affords a complete verification of the formulæ which have been established.

PROBLEM 2.—It is required to *quadrisect* an ellipsoid, the plane of contact between the two frusta being drawn in any direction through the centre.

Let $V_{,}$ V_{2} be the volumes of the two central blocks, V_3 V_4 those of the segments; then, for determining the former, by writing $p_2 \ = \ 0$ in (X), (2), Cor. 3, we have,

$$V_{,} \ = \ \frac{\pi \, abc}{3 \, p_{,}^{3}} \, (3 p_{,}^{2} \ - \ p^{2}) \, p \ = \ V_{2} \ :$$

and, for the segments,

$$V_3 \ = \ \frac{\pi \, abc}{3 \, p_{,}^{3}} \, (p_{,} \ - \ p)^{2} \, (2 \, p_{,} \ + \ p) \ = \ V_4 \ :$$

consequently, for the quadrisection,

$$\frac{2 \, \pi \, abc}{3 \, p_{,}^{3}} \, (3 \, p_{,}^{2} \ - \ p^{2}) \, p = \frac{2 \, \pi \, abc}{3 \, p_{,}^{3}} \, (p_{,} \ - \ p)^{2} \, (2 \, p_{,} \ + \ p).$$

Hence, $$p^3 \ - \ 3 p_{,}^{2} p \ + \ p_{,}^{3} \ = \ 0.$$

or, $$\cos^3 \theta \ - \ 3 \cos \theta \ + \ 1 \ = \ 0.$$

The equation for θ or p has three real roots, two of which only are admissible, and the solution will give, for the position of the dividing planes,

$$p \ = \ 0.3473 \, p_{,}.$$

PROBLEM 3.—It is required to determine the volume of any Cone, of which the vertex coincides with the centre of a given ellipsoid, and which is bounded by that surface.

Let V be the volume of the solid which it is the object of the question to determine; then,

$$V = V_, + V_2.$$

Let a plane elliptic section, anywhere situated, be the base of the Cone, $V_,$ its volume. If A is the area of the elliptic base, it has been shewn in (VII), (14), that,

$$A = \frac{\pi\, abc}{p_,^3} (p_,^2 - p^2).$$

From the known expression for the volume of a cone, which has a plane base,

$$V_, = \frac{1}{3} A\, p ;$$

$$\therefore \quad V_, = \frac{\pi\, abc}{3\, p_,^3} (p_,^2 - p^2)\, p.$$

Again, if V_2 is the volume of the segment, contained between the elliptic base and the surface, we find from (X), (3),

$$V_2 = \frac{\pi\, abc}{3\, p_,^3} (p_, - p)^2 (2\, p_, + p).$$

Hence, after the reductions requisite, we find,

$$V = \frac{2\, \pi\, abc}{3} \frac{p_, - p}{p_,}.$$

Cor. 1.—If $p = 0$, $V = \dfrac{2\, \pi\, abc}{3}$, as it should be.

If $p = - p_,$, $V = \dfrac{4\, \pi\, abc}{3}$;

so that this problem affords a determination for the volume of an ellipsoid.

Cor. 2.—Let V' be the volume of the semi-ellipsoid, then,

$$V : V' :: p_, - p : p_,$$

which cannot but be regarded as a singular and beautiful relation.

v

PROBLEM 4.—It is required to determine the volume of the solid, which is cut from a given ellipsoid by a Cylinder, of which the axis passes through the centre in any direction.

Let each of the two equal and opposite plane elliptic sections be represented by A; then, if $V_{\prime}$ is the volume included between these limiting sections and the lateral surface of the cylinder,

$$V_{\prime} \ = \ 2\,A\,p.$$

Let V_{2} be the volume of the segments, intercepted between the two terminal elliptic planes and the surface;

$$\therefore \ V_{2} \ = \ \frac{2\,\pi\,abc}{3\,p_{\prime}^{3}}\,(p_{\prime} - p)^2\,(2\,p_{\prime} + p).$$

If V is the volume which it is required to determine, including the elliptic cylinder, together with the two equal ellipsoidal segments,

$$V \ = \ V_{\prime} \ + \ V_{2}\,;$$

After the reductions requisite, it will be found that,

$$V \ = \ \frac{4\,\pi\,abc}{3}\,\frac{p_{\prime}^{3} - p^{3}}{p_{\prime}^{3}}.$$

COR. 1.—If $\underline{V}$ is the volume of the whole ellipsoid, the last formula will give the singular relation,

$$V \ : \ \underline{V} \ :: \ p_{\prime}^{3} - p^{3} \ : \ p_{\prime}^{3}.$$

COR. 2.—A being the area of the elliptic base,

$$V \ = \ \frac{4}{3}\,\frac{p_{\prime}^{3} - p^{3}}{p_{\prime}^{2} - p^{3}}\,\times\,A.$$

If $p = 0$, in this last expression, the ellipsoid is enveloped by the cylinder, then

$$V \ = \ \frac{2}{3}\,(2\,p_{\prime}\,A).$$

consequently, as in the sphere, *the ellipsoid is equal in volume to $\frac{2}{3}$ of any circumscribing cylinder.*

COR. 3.—From this it appears, that the *volume of a cylinder, in any position, circumscribing the ellipsoid,* is *constant,* and equal to $\frac{3}{2}$ the volume contained by the circumscribed surface.

(XII.) To determine the MASS of any segment, when the density of each parallel section varies as a given fraction of its distance from the centre of the ellipsoid.

If ρ and $d\,\mathrm{M}$ are the density and mass of an infinitessimal section Λ, then, from (X), page 140, $d\,\mathrm{M} = \rho\,d\,\mathrm{V}$;

$$\therefore \quad \mathrm{M} + \mathrm{C} = \frac{\pi\,abc}{3\,p_i{}^3} \int (p_i{}^2 - p^2)\,\rho\,d\,p. \qquad (1)$$

COR.—Let μ be the co-efficient of density, and suppose the density of each parallel section to vary as p^n ; then we have,

$$\mathrm{M} + \mathrm{C} = \frac{\pi\,abc\,\mu}{p_i{}^3} \frac{(n + 3)\,p_i{}^2 - (n + 1)\,p^2}{(n + 1)\,(n + 3)}\,p^{n+1}. \qquad (2)$$

(XIII.) The formulæ which have been established in this chapter suffice to determine many questions of interest relating to the Attraction of an Ellipsoid, upon a particle situated externally or internally; several of which will be now considered.

(XIV.) It is required to ascertain the attraction of a plane lamina, of very small thickness and uniform density, upon a particle anywhere situated ; the attractive force being directly proportional to the mass of the attracting elementary molecules, and to their distance from the particle attracted.

Let t and ρ be the thickness and density of the lamina.

Let any point O, figure 5, be taken as origin ; m the particle attracted ; $m\,\mathrm{Q}$ a perpendicular upon the plane.

Take $d\,\mathrm{A}$ as any elementary area at a point P ; $d\,\mathrm{F}$ the attraction at $d\,\mathrm{A}$ on m in the direction $m\,\mathrm{Q}$.

The attraction at dA in the direction mP is,

$$\text{mass} \times m\,\text{P} = \rho\, t\, d\,\text{A} \times m\,\text{P};$$

and, after this is resolved at right angles to the plane, writing $m\,\text{Q} = k$.

$$d\,\text{F} = \rho\, t\, d\,\text{A} \times m\,\text{Q} = k \rho\, t\, d\,\text{A};$$

Integrating between given limits, we have, therefore,

$$\text{F} = k\,\rho\, t \times \text{A}.$$

If M is the mass of A, $\text{M} = \rho\, t\,\text{A};$

$$\therefore \quad \text{F} = \text{M}\,k. \tag{1}$$

Hence it appears that the attraction towards the lamina is constant, wherever the particle attracted may be situated in a plane parallel to the lamina. The attraction is, further, the same as if the particle were placed in the vertical passing through any point O, figure 5, and attracted by the whole mass of the lamina, supposed to be concentrated in that point.

(XV.) It is required to determine the attraction of any segment of an Ellipsoid, upon a particle anywhere situated; when the attractive force varies directly as the mass of each plane section parallel to the base, and directly as the distance of that section from the point of attraction.

Let A be the elliptic area of any plane section;

$\quad p$ the perpendicular drawn to it from the centre;

$\quad p_{\prime}$ the perpendicular upon a tangent plane parallel to the section.

Then, $\qquad \text{A} = \dfrac{\pi\, abc}{p_{\prime}^{3}} \left(p_{\prime}^{2} - p^{2}\right).$

Let h be the perpendicular drawn from the centre upon a plane which includes the particle attracted and is parallel to the section A; k a perpendicular from the particle upon the section: so that,

$$k = h - p.$$

consequently, $\qquad d\,\text{F} = \dfrac{\pi\, abc}{p_{\prime}^{3}} (h-p)(p_{\prime}^{2}-p^{2})\,\rho\, d p;$

$$\text{F} + \text{C} = \dfrac{\pi\, abc}{p_{\prime}^{3}} \int (h-p)(p_{\prime}^{2}-p^{2})\,\rho\, d p. \tag{2}$$

(XVI.) An ellipsoid, of *uniform density*, attracts a particle anywhere situated ; the attractive force varying with the distance of the attracted particle from each parallel section : the position of the plane sections in regard to direction being entirely arbitrary. It is required to ascertain the attraction of a segment or frustum of the solid.

Since the density is constant, we find from (2), (XV) ;

$$F + C = \frac{\pi \rho \, abc}{12 \, p_i^3} \left\{ 4 \, h \, p \, (3 \, p_i^2 - p^2) - 3 \, p^2 \, (2 \, p_i^2 - p^2) \right\}. \quad (1)$$

Cor.—When the integral (1) is taken between the limits $- p_i$ and $+ p_i$, it will be found that,

$$F = \frac{4 \, \pi \rho \, abc}{3} \, h. \quad (2)$$

It appears, therefore, that the attraction is the same, according to this law, as though the whole solid were condensed into a particle of equal mass at its centre.

(XVII.) The variable section of the surface may be supposed to move parallel to a tangent plane which includes the attracted particle ; and this can take place in an infinite variety of ways. It is required to ascertain the attraction, under these circumstances.

(1.)—Let the limiting section be central ; then, from (1), (XVI),

$$F = \frac{\pi \rho \, abc}{12 \, p_i^3} \left\{ 4 \, h \, p \, (3 \, p_i^2 - p^2) - 3 \, p^2 \, (2 \, p_i^2 - p^2) \right\}. \quad (1)$$

If we consider the attraction of half the solid, the second limit of p is p_i ; and, by the hypothesis, we have, also, $h = p_i$:

$$\therefore \quad F = \frac{5 \, \pi \rho \, abc}{12} \, p_i.$$

(2.)—From (XIV), page 147, it appears that the supposition here made is equivalent to that of placing the particle attracted upon the surface ; and we have for a particle so situated, when,

$p_{,} = a$;   *Greatest attraction* $= \dfrac{5\pi\rho a^2 bc}{12}$ on the *major* axis.

$p_{,} = b$;   *Mean*   „ $= \dfrac{5\pi\rho ab^2 c}{12}$ „   *mean*   „

$p_{,} = c$;   *Least*   „ $= \dfrac{5\pi\rho abc^2}{12}$ „   *least*   „

$\therefore$ G   :   M   :   L   :: a : b : c.

(XVIII.) In the preceding sections the attraction of the semi-ellipsoid which is *nearest* to the particle has been considered; it is required to ascertain the attraction of the remoter half of the solid.

In order to this the integral (1), (XVI), must be corrected between the limits $- p_{,}$, 0. Hence, if F_2 is the attraction of the *remoter* half solid, and $F_{,}$ that of the *nearer* half,

$$F_{,} = \frac{\pi\rho abc}{12}(8h - 3p_{,}).   F_2 = \frac{\pi\rho abc}{12}(8h + 3p_{,}).$$

Cor. 1.—If the particle is placed anywhere *on* the surface, $h = p_{,}$; then,

$$F_{,} = \frac{5\pi\rho abc}{12}p_{,}.   F_2 = \frac{11\pi\rho abc}{12}p_{,}.$$

$$\therefore   F_{,}   :   F_2   ::   5   :   11.$$

Cor. 2.—From the general expressions for $F_{,}$ and F_2 it appears that, in any half of the ellipsoid, the *attraction is neutralised* in a plane at the distance $h = \dfrac{3}{8}p_{,}$ from the centre; $p_{,}$ being the perpendicular upon a tangent plane drawn parallel to the base of the semi-ellipsoid.

(XIX.) It is required to ascertain the attraction of the eighth part of an ellipsoid, bounded by the principal planes, on a particle anywhere situated upon one of the principal axes.

The mass of a section parallel to one of the limiting planes is $=$
$\frac{\pi \rho\, a b c}{4\, p_i^{3}} (p_i^{2} - p^{2})$; in which p_i has either of the values a, b, c :

$$\therefore \quad d\,F \;=\; \frac{\pi \rho\, abc}{4\, p_i^{3}} (h - p)(p_i^{2} - p^{2})\, dp.$$

The integration from $p = 0$ to $p = p_i$ will give,

$$F \;=\; \frac{\pi \rho\, a b c}{48} (8h - 3p_i).$$

By placing the particle, alternately, at each of the three vertices, and denoting by $F_a\ F_b\ F_c$ the corresponding attractive forces, we find,

$$F_a = \frac{5\,\pi \rho\, a^2 b c}{48}. \quad F_b = \frac{5\,\pi \rho\, a b^2 c}{48}. \quad F_c = \frac{5\,\pi \rho\, a b c^2}{48}.$$

(XX.) It is required to ascertain the resultant attraction of an ellipsoid octant upon a particle anywhere situated.

Let $F_a\ F_b\ F_c$ be the attractions towards the limiting planes.

Let $\alpha\,\beta\,\gamma$ be the co-ordinates of the particle, then, by (XIX),

$$F_a = \frac{\pi \rho\, a b c}{48}(8a - 3\alpha). \quad F_b = \frac{\pi \rho\, a b c}{48}(8\beta - 3b).$$

$$F_c = \frac{\pi \rho\, a b c}{48}(8\gamma - 3c).$$

Hence, if F represents the resultant attractive force, the three attractions being imagined to operate simultaneously,

$$F = \frac{\pi \rho\, a b c}{48}\Big\{ (8\alpha-3a)^2 + (8\beta-3b)^2 + (8\gamma-3c)^2 \Big\}^{\frac{1}{2}}.$$

Cor. 1.—If $\alpha\,\beta\,\gamma$ are all negative,

$$F = \frac{\pi \rho\, a b c}{48}\Big\{ (8\alpha+3a)^2 + (8\beta+3b)^2 + (8\gamma+3c)^2 \Big\}^{\frac{1}{2}}.$$

Hence it appears that, for constant values of $\alpha\,\beta\,\gamma$, the attractive force upon the particle is greatest when it is placed in this position; and, generally, it has greater intensity when the particle is placed towards either of the plane faces of the solid.

COR. 2.—If $\alpha\beta\gamma$ are respectively $= 0$, the attracted particle is placed at the angle of the octant; in this case the resultant attractive force is,

$$F = \frac{\pi \rho\, a\, b\, c}{16} (a^2 + b^2 + c^2)^{\frac{1}{2}}.$$

COR. 3.—The attraction is zero when,

$$\alpha = \frac{3}{8} a; \qquad \beta = \frac{3}{8} b; \qquad \gamma = \frac{3}{8} c;$$

but it may be shewn that these co-ordinates indicate the centre of gravity of the octant: when, therefore, a particle is placed at that point, the attraction upon it is neutralised.

COR. 4.—Let $\alpha_i\, \beta_i\, \gamma_i$ be co-ordinates of the particle, measured from the centre of gravity; then,

$$F = \frac{\pi \rho\, a\, b\, c}{6} (\alpha_i^2 + \beta_i^2 + \gamma_i^2)^{\frac{1}{2}}.$$

$$= \frac{1}{8} \left(\frac{4\pi\rho\, abc}{3} \right) (\alpha_i^2 + \beta_i^2 + \gamma_i^2)^{\frac{1}{2}}.$$

$$= \textit{mass of the solid} \times \textit{distance of the attracted particle from the centre of gravity.}$$

The solid, therefore, attracts with the same force as though it were condensed into a molecule at the point of neutral attraction.

COR. 5.—Let δ be $\frac{1}{8}$th the distance of the attracted particle from the point of neutral attraction; $\theta\, \phi\, \psi$ the inclinations of this line to the principal axes: then, $\delta = \Sigma^{\frac{1}{2}}(8a - 3a)^2$, and we find,

$$\cos \theta = \frac{8a - 3a}{\delta}. \qquad \cos \phi = \frac{8\beta - 3b}{\delta}. \qquad \cos \psi = \frac{8\gamma - 3c}{\delta}.$$

These expressions determine the line of action of the resultant attraction.

W. F. MATHEW, Printer, 59, St. George Street, Cape Town.

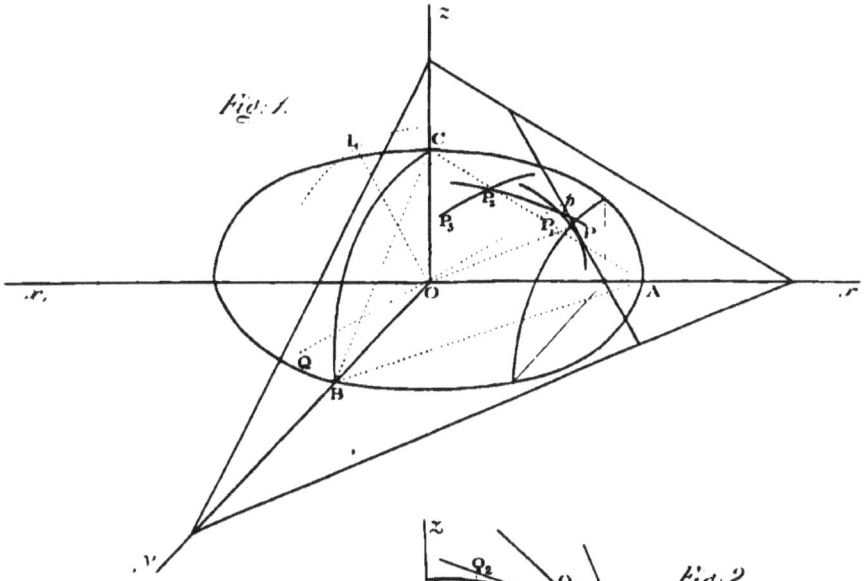

Fig. 1.

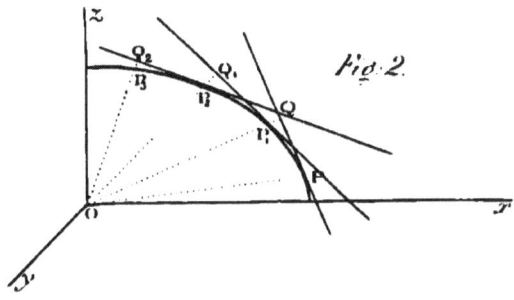

Fig. 2.

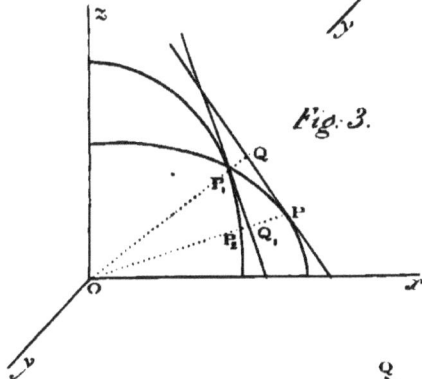

Fig. 3.

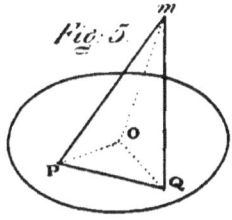

Fig. 5.

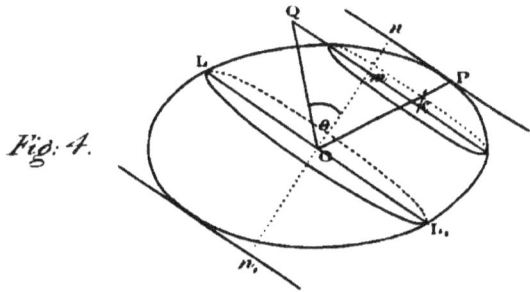

Fig. 4.

www.ingramcontent.com/pod-product-compliance
Lightning Source LLC
Chambersburg PA
CBHW020905210326
41598CB00018B/1773